"This book is a treasure trove of insights for anyone considering emerging robotic systems in their context: in which they operate, are designed, and imagined. A stellar team of editors and authors offer a well-structured overview of established and emerging design methodologies, which challenged my perspectives on robotic capabilities, human-robot ecologies, and narratives. The second half of the book provides important essays that discuss challenges around robotic innovation and design. If you want to broaden your perspectives around what robotic capabilities are and should be: this rich book is for you."

— **Prof. dr. ir. David Abbink**, *Delft University of Technology, Scientific Director of FRAIM, the Netherlands*

"This book presents robots not as a sci-fi trope but as a realm of design possibility. It will be instrumental to helping to reshape tomorrow's robots so that they help fulfill human needs and desires. It is rich with viewpoints and novel perspectives which help to equip designers with what they need to build groundbreaking robots."

— **Wendy Ju**, *Jacobs Technion-Cornell Institute at Cornell Tech, USA*

"Whenever we embark on the development of innovative robotic applications working closely with people in our daily lives, we're faced with the intricate challenge of crafting interactions for these state-of-the-art creations. In this complex endeavor, no single design choice reigns supreme. This book provides ideas about diverse design choices to explore and important aspects we should carefully think during the design process. I highly recommend this book for students and researchers who are interested in the design of such interactions."

— **Takayuki Kanda**, *Graduate School of Informatics, Kyoto University, Japan*

"Good design often goes unnoticed, but good design methods deserve our full attention. This book packs expertise from various fields into an exciting guide on how to design robots and their interaction with people, perfect for robot builders, students, and researchers."

— **Tony Belpaeme**, *Ghent University, Belgium*

Designing Interactions with Robots

Developing robots to interact with humans is a complex interdisciplinary effort. While engineering and social science perspectives on designing human–robot interactions (HRI) are readily available, the body of knowledge and practices related to design, specifically interaction design, often remain tacit. ***Designing Interactions with Robots*** fills an important resource gap in the HRI community, and acts as a guide to navigating design-specific methods, tools, and techniques.

With contributions from the field's leading experts and rising pioneers, this collection presents state-of-the-art knowledge and a range of design methods, tools, and techniques, which cover the various phases of an HRI project. This book is accessible to an interdisciplinary audience, and does not assume any design knowledge. It provides actionable resources whose efficacy have been tested and proven in existing research.

This manual is essential for HRI design students, researchers, and practitioners alike. It offers crucial guidance for the processes involved in robot and HRI design, marking a significant stride toward advancing the HRI landscape.

Chapman & Hall/CRC Artificial Intelligence and Robotics Series
Series Editor: Roman Yampolskiy

Explainable Agency in Artificial Intelligence
Research and Practice
Silvia Tulli and David W. Aha

An Introduction to Universal Artificial Intelligence
Marcus Hutter, Elliot Catt and David Quarel

AI: Unpredictable, Unexplainable, Uncontrollable
Roman V. Yampolskiy

Transcending Imagination
Artificial Intelligence and the Future of Creativity
Alexander Manu

Responsible Use of AI in Military Systems
Jan Maarten Schraagen

AI iQ for a Human-Focused Future
Strategy, Talent, and Culture
Seth Dobrin

Federated Learning
Unlocking the Power of Collaborative Intelligence
Edited by M. Irfan Uddin and Wali Khan Mashwan

Energy Efficiency and Robustness of Advanced Machine Learning Architectures
A Cross-Layer Approach
Alberto Marchisio and Muhammad Shafique

Designing Interactions with Robots
Methods and Perspectives
Edited by Maria Luce Lupetti, Cristina Zaga, Nazli Cila, Selma Šabanović, and Malte F. Jung

For more information about this series please visit: https://www.routledge.com/Chapman--HallCRC-Artificial-Intelligence-and-Robotics-Series/book-series/ARTILRO

Designing Interactions with Robots

Methods and Perspectives

Edited by
Maria Luce Lupetti, Cristina Zaga, Nazli Cila, Selma Šabanović, and Malte F. Jung

Designed cover image: Mafalda Gamboa

First edition published 2025
by CRC Press
2385 NW Executive Center Drive, Suite 320, Boca Raton FL 33431

and by CRC Press
4 Park Square, Milton Park, Abingdon, Oxon, OX14 4RN

CRC Press is an imprint of Taylor & Francis Group, LLC

© 2025 selection and editorial matter, Maria Luce Lupetti, Cristina Zaga, Nazli Cila, Selma Šabanović, and Malte F. Jung; individual chapters, the contributors

Reasonable efforts have been made to publish reliable data and information, but the author and publisher cannot assume responsibility for the validity of all materials or the consequences of their use. The authors and publishers have attempted to trace the copyright holders of all material reproduced in this publication and apologize to copyright holders if permission to publish in this form has not been obtained. If any copyright material has not been acknowledged please write and let us know so we may rectify in any future reprint.

The Open Access version of this book, available at www.taylorfrancis.com, has been made available under a Creative Commons Attribution-NonCommercial-NoDerivatives (CC-BY-NC-ND) 4.0 International license.

Funding is from: TU Delft Open Access Fund; National Science Foundation under Grant No. 1942085; University of Twente; Indiana University Bloomington; Unite! Seed Funding – Roboethic – Gratz University of Technology.

Trademark notice: Product or corporate names may be trademarks or registered trademarks and are used only for identification and explanation without intent to infringe.

ISBN: 978-1-032-44212-9 (hbk)
ISBN: 978-1-032-43027-0 (pbk)
ISBN: 978-1-003-37102-1 (ebk)

DOI: 10.1201/9781003371021

Contents

About the Editors ... x
List of Contributors ... xii
Preface ... xv

Chapter 1 Introduction ... 1
Maria Luce Lupetti

Chapter 2 Designing Artificial Agents: Appearance and Expressivity ... 6
Maria Luce Lupetti

Box 2.1 The Robotization Design of Everyday Things ... 11
Sonya S. Kwak and Dahyun Kang

Box 2.2 Ready Made Prototyping ... 14
Ilan Mandel and Wendy Ju

Box 2.3 Multi-modal Expressivity ... 17
Kim Baraka

Box 2.4 Minimalism ... 20
Oren Zuckerman and Hadas Erel

Box 2.5 Animation ... 23
Tiago Ribeiro

Box 2.6 Bodystorming ... 26
Danilo Gallo, Maria Antonietta Grasso, Kahyeon Kim, and Sure Bak

Box 2.7 Dramaturgy for Devices ... 29
Maaike Bleeker and Marco C. Rozendaal

Box 2.8 Worldbuilding ... 32
Christine P. Lee, Bengisu Cagiltay, and Bilge Mutlu

Chapter 3 Designing for Social Embeddedness: Mutually Shaping Robots and Society ... 38
Selma Šabanović

Box 3.1 User Enactments ... 43
Jodi Forlizzi, Carl DiSalvo, Bilge Mutlu, Min Kyung Lee, Michal Luria, Samantha Reig, and John Zimmerman

Box 3.2 Robot Value Mapping....46
Irene Gonzalez and Jan Jacobs

Box 3.3 Collaborative Map-Making....49
Hee Rin Lee

Box 3.4 Pretotyping....52
Ioana Ocnarescu and Isabelle Cossin

Box 3.5 Co-design....55
Katie Winkle

Box 3.6 Breaching Experiments....58
Mads Bødker and Rosenthal-von Der Pütten

Box 3.7 End-User Programming....61
Emmanuel Senft

Box 3.8 Installations and Performances....64
Mari Velonaki

Chapter 4 Designing Human–Robot Ecologies: Beyond Utilitarian Relations....70

Nazli Cila

Box 4.1 Conversations with Agents....74
Iohanna Nicenboim and Elisa Giaccardi

Box 4.2 Techno-Mimesis....77
Judith Dörrenbächer

Box 4.3 Objects with Intent....80
Marco C. Rozendaal

Box 4.4 Aesthetics of Friction....83
Matthias Laschke and Marc Hassenzahl

Box 4.5 Symbiosis....86
Michio Okada

Box 4.6 Playfulness....89
Marius Hoggenmueller

Box 4.7 Para-functionality....92
James Pierce

Box 4.8 Relationality....95
Dave Murray-Rust

Chapter 5 Designing Robotic Imaginaries: Narratives and Futures.................100

Cristina Zaga

Box 5.1 Futuristic Autobiographies..105
EunJeong Cheon and Norman Makoto Su

Box 5.2 Soma Design..108
Mafalda Gamboa and Joseph La Delfa

Box 5.3 Metaphors..111
Patricia Alves-Oliveira

Box 5.4 Design Fiction..114
Simone Rebaudengo

Box 5.5 Speculative Design..117
James Auger

Box 5.6 Adversarial Design..120
Maria Luce Lupetti

Box 5.7 Ethnographic Experiential Futures..123
Michal Luria

Box 5.8 Blending Traditions..126
Haipeng Mi

Chapter 6 Choosing Materials for Personal Robot Design............................132

Guy Hoffman

Chapter 7 Designing Robots that Work and Matter............................140

Carla Diana

Chapter 8 Critical Perspectives in Human–Robot Interaction Design.............148

Sara Ljungblad and Mafalda Gamboa

Chapter 9 Understanding Designerly Contributions............................161

Nazli Cila

Chapter 10 Toward a Future Beyond Disciplinary Divides............................170

Cristina Zaga

Index..182

Editors

Maria Luce Lupetti is an Assistant Professor in Interaction and Critical Design at the Department of Architecture and Design at Politecnico di Torino (IT). Her research is concerned with all matters of human entanglement with the artificial world, especially concerning complex technologies such as AI and robotics. She also serves as *Exhibit X* section editor for *Interactions* Mag. She is a former core member of the AiTech Initiative on Meaningful Human Control over AI Systems and of the Automated Mobility Lab, at TU Delft, where she worked for several years. Maria Luce Lupetti received a PhD cum Laude in Production, Management and Design from Politecnico di Torino, in Italy, for her research through design investigations into the field of educational robotics for children.

Dr. Cristina Zaga is an assistant professor of Human-Centered Design group and DesignLab at the University of Twente (NL). Cristina's research aims to foster societal transitions toward justice, care, and solidarity, with a focus on the future of work and care with robots and AI. They lead the Social Justice and AI networks, working toward mitigating the dehumanizing effects of AI and promoting social and environmental justice. Their award-winning work has received many accolades, including the NWO Science Prize for DEI initiatives (2022), the Dutch High Education Award (2022), and the Google Women Techmaker Award and scholarship (2018).

Nazli Cila is an Assistant Professor at the Department of Human-Centered Design at Delft University of Technology (NL). Her work seeks to understand how to design symbiotic relationships between humans and AI while preserving individual, ethical, and societal values, such as autonomy, enrichment, and justice. In addition to investigating design qualities and societal implications of human-AI collaborations, she is also fascinated by how designers and design researchers produce knowledge. This occasionally zeros in on studying the complex landscape of robot design, at times, expands its focus to embrace design epistemology and methodology at a broader level. She is in the steering committee of the TU Delft AI Labs program and the co-director of the AI DeMoS Lab, investigating how to facilitate responsible design and use of AI for a meaningful democratic engagement.

Selma Šabanović is Professor of Informatics and Cognitive Science at Indiana University Bloomington. She studies social robotics and human–robot interaction, with a focus on exploring how robots should be designed to assist people in various use contexts, including mental health, wellness, education, and social participation. She works with current and potential robot users of all ages, from children to older adults, and in various cultures, including East Asia, Europe, and the US. She served as the Editor in Chief of the ACM Transactions on Human–Robot Interaction from 2017 to 2024, and currently serves as an Associate Vice President of the IEEE Robotics and Automation Society Educational Activities Board. She received her PhD in Science and Technology Studies in 2007 from Rensselaer Polytechnic Institute.

Malte F. Jung is an Associate Professor in the Information Science Department at Cornell University and the Nancy H. '62 and Philip M. '62 Young Sesquicentennial Faculty Fellow. His research brings together approaches from design and behavioral science to build an understanding about how we can build robots that function better in group and team settings. His work has received several awards, including an NSF CAREER award. He holds a Ph.D. in Mechanical Engineering, and a PhD Minor in Psychology from Stanford University, and a Diploma in Mechanical Engineering from the Technical University of Munich. Prior to joining Cornell, Malte Jung completed a postdoc at the Center for Work, Technology, and Organization at Stanford University.

Contributors

Patricia Alves-Oliveira
University of Michigan
Ann Arbor, Michigan

James Auger
École normale supérieure Paris-Saclay
Paris, France

Sure Bak
Gyeonggi-do, Naver Labs
South Korea

Kim Baraka
Vrije Universiteit Amsterdam
Amsterdam, Netherlands

Maaike Bleeker
Utrecht University
Utrecht, Netherlands

Mads Bødker
Copenhagen Business School
Frederiksberg, Denmark

Bengisu Cagiltay
University of Wisconsin–Madison
Madison, Wisconsin

EunJeong Cheon
Syracuse University
Syracuse, New York

Nazli Cila
Delft University of Technology
Delft, Netherlands

Isabelle Cossin
STRATE
Sèvres, France

Dahyun Kang
Korea Institute of Science and Technology
Seoul, South Korea

Carla Diana
Cranbrook Academy of Art
Bloomfield Hills, Michigan

Carl DiSalvo
Georgia Institute of Technology
Atlanta, Georgia

Judith Dörrenbächer
University of Siegen
Siegen, Germany

Hadas Erel
Reichman University
Herzliya, Israel

Jodi Forlizzi
Carnegie Mellon University
Pittsburgh, Pennsylvania

Danilo Gallo
Naver Labs
Grenoble, France

Mafalda Gamboa
Chalmers University of Technology
Gothenburg, Sweden

Elisa Giaccardi
Politecnico di Milano
Milan, Italy

Irene Gonzalez
AppFolio
Santa Barbara, California

Maria Antonietta Grasso
Naver Labs
Grenoble, France

Marc Hassenzahl
University of Siegen
Siegen, Germany

Guy Hoffman
Cornell University
Ithaca, New York

Marius Hoggenmueller
University of Sydney
Camperdown, Australia

Jan Jacobs
Lely
Maassluis, Netherlands

Wendy Ju
Cornell University
Ithaca, New York

Kahyeon Kim
Naver Lab
Gyeonggi-do, South Korea

Sonya S. Kwak
Korea Institute of Science and Technology
Seoul, South Korea

Joseph La Delfa
KTH Royal Institute of Technology
Stockholm, Sweden

Matthias Laschke
University of Siegen
Siegen, Germany

Christine P. Lee
University of Wisconsin–Madison
Madison, Wisconsin

Hee Rin Lee
Michigan State University
East Lansing, Michigan

Min Kyung Lee
University of Texas
Austin, Texas

Sara Ljungblad
Chalmers University of Technology
Gothenburg, Sweden

Maria Luce Lupetti
Politecnico di Torino
Turin, Italy

Michal Luria
Center for Democracy and Technology
Washington, District of Columbia

Ilan Mandel
Cornell University
Ithaca, New York

Haipeng Mi
Tsinghua University
Beijing, China

Dave Murray-Rust
Delft University of Technology
Delft, Netherlands

Bilge Mutlu
University of Wisconsin-Madison
Madison, Wisconsin

Iohanna Nicenboim
Delft University of Technology
Delft, Netherlands

Ioana Ocnarescu
STRATE
Sèvres, France

Michio Okada
Toyohashi University of Technology
Toyohashi, Japan

James Pierce
University of Washington
Seattle, Washington

Simone Rebaudengo
OIO Studio
London, United Kingdom

Samantha Reig
University of Massachusetts Lowell
Lowell, Massachusetts

Tiago Ribeiro
Soul Machines
Auckland, New Zealand

Marco C. Rozendaal
Delft University of Technology
Delt, Netherlands

Emmanuel Senft
Idiap Research Institute
Martigny, Switzerland

Selma Šabanović
Indiana University
Bloomington, Indiana

Makoto Norman Su
University of California
Los Angeles, California

Mari Velonaki
University of New South Wales
Kensington, Australia

Katie Winkle
Uppsala Universitet
Uppsala, Sweden

Cristina Zaga
University of Twente
Enschede, Netherlands

John Zimmerman
Carnegie Mellon University
Pittsburgh, Pennsylvania

Oren Zuckerman
Reichman University
Herzliya, Israel

Preface

Malte F. Jung This book was born out of our frustration with the state of design in human–robot interaction. As researchers, we were disheartened by our community's fixation on research approaches borrowed from psychology and the behavioral sciences. We observed our students and colleagues engage in inspiring design projects that push the boundaries of HRI, only to see their innovative work constrained by the conventional format of a "user study." This often resulted in the loss of nuanced insights and learnings derived from their design practice. Why do we need bar-charts and p-values to legitimize design work as "scientific" enough for our top conferences and journals?

As designers, we were disconcerted to see so many of the new machines that increasingly inhabit our lives be designed with so little regard for people and the world we inhabit. The prevalent "technology-first" approach overlooks the complexities of human needs, our environmental contexts, and the intricate social ecologies into which we introduce our robots. Why is there such a pervasive inclination toward technology-driven design, rather than adopting a view that prioritizes understanding people, their environments, and the social dynamics at play?

In a way, the work on this book began with a workshop on "Designerly HRI Knowledge," led by Maria Lupetti, Cristina Zaga, and Cila Nazli and held at the International Symposium on Robot and Human Interactive Communication in 2020. A position paper on "Designerly ways of knowing in HRI" soon followed a year later. It quickly became apparent that the time is right for the HRI community to engage more deeply with design as an important topic and with designing as a legitimate form of research. That's when we came together to start this book.

The aim of our book is to showcase and celebrate ways of designing and designerly research that are already happening in HRI. Browsing this book, we hope you will find that design is alive and well in HRI. We hope this work inspires many others to incorporate design practices into their work, to further establish designing as a legitimate research practice in HRI, and to work toward a desirable future with robots.

1 Introduction

Maria Luce Lupetti
Politecnico di Torino, Turin, Italy

There are few artificial things in the world able to catch our interest as robots do. Often, a robot doesn't even need to be moving or doing things to make us curious – it can be enough that we are told that it is a robot. We touch, push, poke, sometimes name and dress them. But, *why? What is in a robot that grabs our attention and encourages us to do (sometimes) irrational things?* Understanding this invites us to foremost look at what a robot is in the first place. The answer is all but obvious.

Many have attempted to define what a robot is, and no univocal conceptualizations exist. Early definitions emphasize the mechanical and utilitarian nature of this artifact, such as the classical one by the Robot Institute of America, describing a robot as "*a reprogrammable, multifunctional manipulator designed to move material, parts, tools, or specialized devices through various programmed motions for the performance of a variety of task* (Considine and Considine, 1986)". In contrast, other examples often highlight the resemblance to living things, either human or zoomorphic, as characterizing features, like the one by Thrun (2004), who describes a robot as "*an automatic device that performs functions normally ascribed to humans or a machine in the form of a human*".

Whether it is framed as more mechanical or life-like, the term "robot" is quite univocally used to describe an artificial agent intended to act somewhat autonomously for serving and supporting people in various ways, from taking over dangerous, dirty, and dull tasks in working environments (Engelberger, 2012) to providing care and company in social contexts (Breazeal et al., 2016). To this traditional and practical way of conceptualizing robots, however, we suggest looking at the complexity of these agents in relation to our social environments and cultural life. Robots are things that sit in a space between technological progress and folklore, between utility and wonder, and that can hardly be separated from our rhetoric about the artificial, societal ideals of progress, and collective visions of the future.

Robots regularly come embedded in a narrative of social progress (Šabanović, 2010) in which we are liberated from the burdens of life and are entertained in a world of enhanced – even enchanted – social interactions. Furthermore, the collective imaginary we hold about robots and robotic futures isn't one where we will have some sparce units doing hyperspecialized jobs, but rather millions of robots taking over both our work and domestic spheres, as well as our public life, in a multitude of ways. Many media sources present us with this vision of robots becoming as ubiquitous as personal computers are today, from science fiction novels and movies to business forecasts.

Such narratives and imaginary, however, are not only a product of contemporary discourses around robotics. They have rather sedimented over centuries through

DOI: 10.1201/9781003371021-1

This chapter has been made available under a CC-BY-NC-ND license.

scientific literature, but also and foremost, through popular culture and the arts. Since ancient times, from both west and east civilizations, we find evidence of robot precedents. Ancient legends tell the stories of mechanical entities that were serving, entertaining, or challenging people (Longsdon, 1984), such as the one of *Talos*, a giant bronze statue of a man designed for patrolling the island of Crete (Merlet, 2000), or the fictional automata that can be found in *Liezi*, a traditional Taoist Chinese text dated 300 BC (Mavridis, 2015). Furthermore, ancient automata were not just part of myths and stories, they were also physical demonstrations of technological mastery, such as several entertaining machines developed by Al-Jazari (Mavridis, 2015), which included water powered clocks, fountains, perpetual flutes, and many more (Nocks, 2007). Over time, these marvelous machines grew in complexity, both in their imagined and real manifestations, evidencing the inextricable relationship of technological progress with popular culture and the arts.

This book, then, is an invite to engage with this broader framing, one that asks us to look at *robots as artificial agents whose technological sophistication is intrinsically entangled with rhetoric of automation and marvel.*

In doing so, we encourage loosening the disciplinary boundaries that traditionally characterize research and development in the field of robotics and human–robot interaction (HRI), such as mechanical engineering, computer science, and psychology, to favor conversations among diverse epistemologies coming from both technical and humanistic fields, as well as from both applied and theoretical research traditions.

By virtue of this broad framing, we see the book also as a platform for understanding and communicating what HRI has to offer to adjacent fields. For instance, the recently growing community of conversational user interfaces (CUI), whose embodied materializations include popular smart home assistants such as Amazon Alexa or Google Home, to a large extent occupies a similar space to HRI. These agents sit within a rich imaginary of super-human intelligence, ubiquity, and fear of possible unintended consequences – an imaginary crystallized through various media portrayals. CUIs also share with robots the resemblance of human entities, only translated into a narrower set of skills and interactional qualities. Further, aspects of socio-technical imagination, autonomy, and distribution of control are shared between these communities and even more present in the once-unrelated field of mobility. With the recent development in sensing technologies and the fast-paced progress in deep learning algorithms, however, cars are now also increasingly being regarded as robots.

All in all, if we look closely at the socio-technical complexity and rhetorical intricacies of robots, we find many parallels with current discourses around Artificial Intelligence (AI) and its diverse applications. This allows us to see beyond the common view that only stresses the advantages brought by AI for developing robots, to unveil a vast body of knowledge that the robotics and HRI fields have to offer to AI, not only in technical terms but also and foremost regarding matters of social and ethical implications.

Signs of these blurring boundaries between fields and disciplines are increasingly visible in academia, as workshops and special issues dedicated to the adjacent fields of AI, CUI, and autonomous driving (just to name a few) are regularly hosted by the major HRI conferences and journals, and vice versa, we find HRI perspectives in the flagship venues of these fields. This encourages diverse communities to come together and collectively address matters of agency, control, anthropomorphism, social perception, justice, and collective imagination. Each community, from its distinct standpoint, implicitly engages with the need – we mentioned above – of *broadening our definition of what a robot is* and unconsciously (or not) helps *dismantle preconceived ideas we hold about who should be entitled to shape robots and our futures with them.* But the more we argue for broadening our view on robotics and loosening our disciplinary boundaries, the more a question arises. *Why, then, a design book?*

Design clearly stands as an independent discipline (Cross, 2019), with its skills, values, interests (Archer, 1979), and ways of thinking and knowing (Cross, 1982). Yet it is also extensively appreciated and adopted by disparate fields, from sciences to humanities, as a powerful mindset to address complex societal challenges (Mejía et al., 2023). We now find design regularly involved in innovation in the public sector and policy making (Bason, 2016), education (Koh et al., 2015), healthcare (Groeneveld et al., 2018), and tech-driven innovation (Norman and Verganti, 2014), including robotics and HRI. Among the many aspects of design that gather large interest is its distinct commitment with building appropriate conceptions and modeling of the artificial world and how we relate to it (Cross, 1982) – the act of framing (Kolko, 2010) and steering research toward doing the "right thing" as opposed to doing "things right" (Luria et al., 2021).

This comes through a practice of contamination of skills and constant creative exercise of lateral thinking that allows for unusual connections to be drawn and uncharted territories to be explored (Imbesi et al., 2020). Design methods and thinking help us disregard the 'drawers' in which we have been trained (Lindberg et al., 2010) also thanks to its constant appropriation of practices from multiple disciplines, and the translation of theories into actionable principles. We then encourage looking at design as an integrative (Buchanan, 1992), almost undisciplined field (Imbesi et al., 2020), as space of doing and learning that is incessantly "in-between", pushing the borders of knowledge and praxis (Imbesi et al., 2020).

This book looks at the growing presence of design in the field of robotics and HRI, and draws a conceptual map of broad as well as emergent methods and perspectives and their implications, and to ultimately make these practices more accessible to all. Our ambition with this book is to make these practices, once the sole purview of those identifying as 'designers', a resource for all who actually design.

NAVIGATING THE BOOK

The book is structured into two main parts. The first draws a map of popular as well as emerging design practices, which we organized into four main sections each providing a different focus: *designing artificial agents–appearance & expressivity* (Chapter 2); *designing for social embeddedness: mutually shaping robots and society* (Chapter 3); *designing human–robot ecologies: beyond utilitarian relations* (Chapter 4); and *designing robotic imaginaries: narratives & futures* (Chapter 5).

Each of these chapters is accompanied by eight methods and perspectives provided by external authors who were selected among the most prominent researchers and practitioners engaged in the design of robotic technologies.

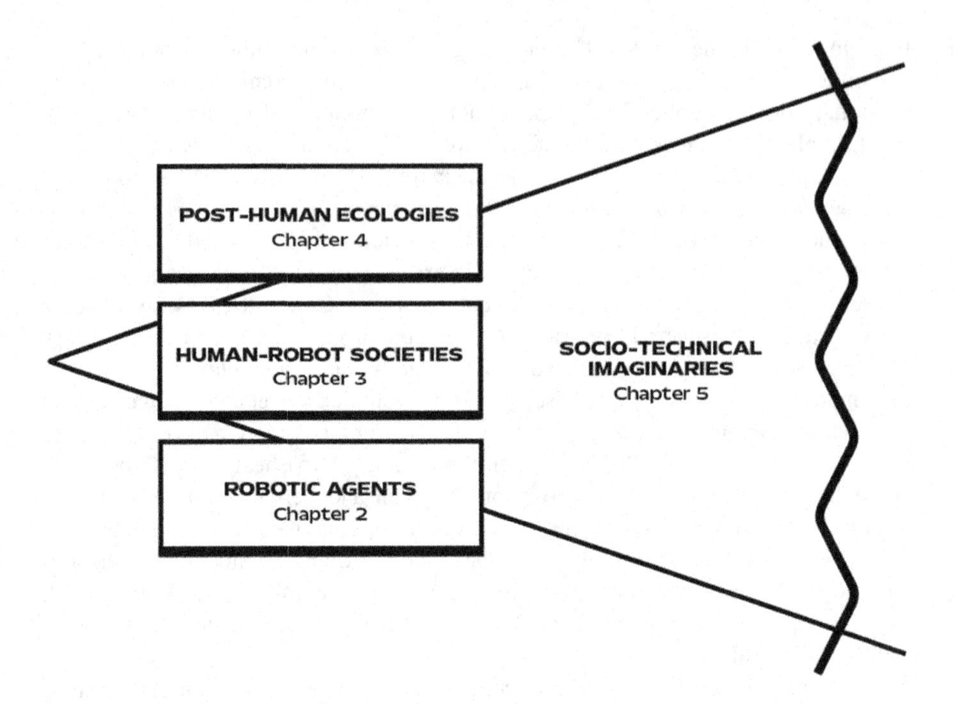

FIGURE 1.1 Overarching structure of the book contents and a vision for the design of robots and HRI: from focusing to an agent's expressivity, to configuring human–robot societies and ecologies beyond human-centeredness, to understanding and shaping broader socio-technical imaginaries. Chapters 3–6, each present a discussion of one of these layers and provide a set of relevant methods and perspectives. The final essays of the book also deepen some aspects of this conceptual structure, but focus more on epistemic and methodological reflections around the design of robots and HRI.

The second part of the book collects essays from external authors as well as the editors, each providing a deep dive into matters at the intersection of design and robotic innovation. In particular, the authors discuss: the distinctive *materiality of robots* (Chapter 6); the *complexity of making robots work in the real world* (Chapter 7); the need for *critical perspectives in robotics* (Chapter 8); the ways in which *design practices can contribute to knowledge about robots and HRI* (Chapter 9); and a final outlook into *how we can move toward a future beyond epistemological divides* (Chapter 10).

REFERENCES

Archer, B. (1979). Design as a discipline. *Design Studies*, *1*(1), 17–20.

Bason, C. (2016). *Design for policy*. Routledge.

Breazeal, C., Dautenhahn, K., & Kanda, T. (2016). Social robotics. *Springer handbook of robotics, 1935–1972*. Springer.

Buchanan, R. (1992). Wicked problems in design thinking. *Design Issues*, *8*(2), 5–21.

Considine, D. M., & Considine, G. D. (1986). Robot technology fundamentals. In *Standard handbook of industrial automation* (pp. 262–320). Springer: US.

Cross, N. (1982). Designerly ways of knowing. *Design Studies*, *3*(4), 221–227.
Cross, N. (2019). Design as a discipline. *Design Studies*, *65*, 1–5.
Engelberger, J. F. (2012). *Robotics in practice: Management and applications of industrial robots*. Springer Science & Business Media.
Groeneveld, B., Dekkers, T., Boon, B., & D'Olivo, P. (2018). Challenges for design researchers in healthcare. *Design for Health*, *2*(2), 305–326.
Imbesi, L., Luca, S., & Canevacci, M. (2020). The categories collapse, knowledge is mobile: Dislocations between design and anthropology. *DIID. Disegno Industriale Industrial Design*, *70*, 34–41.
Koh, J. H. L., Chai, C. S., Wong, B., Hong, H. Y., Koh, J. H. L., Chai, C. S., … & Hong, H. Y. (2015). *Design thinking and education* (pp. 1–15). Springer: Singapore.
Kolko, J. (2010). Sensemaking and framing: A theoretical reflection on perspective in design synthesis.
Lindberg, T., Noweski, C., & Meinel, C. (2010). Evolving discourses on design thinking: How design cognition inspires meta-disciplinary creative collaboration. *Technoetic Arts*, *8*(1), 31–37.
Longsdon, T. (1984). *The robot revolution*. Simon & Schuster Inc.
Luria, M., Hoggenmüller, M., Lee, W. Y., Hespanhol, L., Jung, M., & Forlizzi, J. (2021, March). Research through design approaches in human-robot interaction. In *Companion of the 2021 ACM/IEEE international conference on human-robot interaction* (pp. 685–687).
Mavridis, N. (2015). A review of verbal and non-verbal human–robot interactive communication. *Robotics and Autonomous Systems*, *63*, 22–35.
Mejía, G. M., Henriksen, D., Xie, Y., García-Topete, A., Malina, R. F., & Jung, K. (2023). From researching to making futures: A design mindset for transdisciplinary collaboration. *Interdisciplinary Science Reviews*, *48*(1), 77–108.
Merlet, J. P. (2000). A historical perspective of robotics. In *International symposium on history of machines and mechanisms proceedings HMM 2000* (pp. 379–386). Springer: Netherlands.
Nocks, L. (2007). *The robot: The life story of a technology*. Greenwood Publishing Group.
Norman, D. A., & Verganti, R. (2014). Incremental and radical innovation: Design research vs. technology and meaning change. *Design Issues*, *30*(1), 78–96.
Šabanović, S. (2010). Robots in society, society in robots: Mutual shaping of society and technology as a framework for social robot design. *International Journal of Social Robotics*, *2*(4), 439–450.
Thrun, S. 2004. Toward a framework for human-robot interaction. *Human-Computer Interaction*, *19*(1), 9.

2 Designing Artificial Agents

Appearance and Expressivity

Maria Luce Lupetti
Politecnico di Torino, Turin, Italy

At its very basis, designing robots is about defining their appearance and expressivity, through combinations of shapes, forms, colors, and textures that, altogether, generate a semantics (Krippendorff & Butter, 1984). One could think of appearance and expressivity as futilities since these operate at the surface of the artifact and are not necessarily linked with the underlying functionalities. Yet, the very way an artifact looks and mingles with our wants and desires–beyond fulfilling our needs (Crilly et al., 2004)–is key to achieving understandability, acceptance, and affection. As Crilly et al. (2004) argue, appearance is to a large extent at the basis of our judgment of products; we unconsciously use visual information to assess the aesthetic quality, functionality, and social significance of man-made things.

The semantics of artifacts then, on the one hand, provides practical indications of "what something is and how that something should be handled" (Demirbilek & Sener, 2003), but on the other, it is what turns things into communication devices. More precisely, the perceived attributes composing a specific design elevate a product from being a mere object to becoming a *transmitter* of a message situated within a communication system, involving designers as the source of the message, users as receivers and destination, and the environment providing the channels and conditions for the message to transit (Monö et al., 1997).

Furthermore, understanding both the practical as well as the rhetorical structures emanating from specific configurations is of strategic importance for designers as these strongly affect the visceral and affective reaction we have toward products. People tend to respond to product experiences with a variety of emotions, each carrying different values and implications in terms of perceived novelty, pleasantness, legitimacy, and more (Desmet, 2003). For instance, the therapeutic robot *Paro* is designed to mingle with the fantasy we hold about what interacting with a baby seal would feel like, even though a real interaction with this animal would be rather different (Calo et al., 2011). As for a teddy bear (Morris et al., 1995), the cute and cuddly configuration of the robot is not intended to mimic reality, but to rather stimulate a positive emotional response and nurturing behaviors.

Emotional responses, however, may be more difficult to decipher when less explicitly linked to the external configuration of the product. Industrial robotic arms, for

DOI: 10.1201/9781003371021-2
This chapter has been made available under a CC-BY-NC-ND license.

instance, often raise adverse feelings—even fear (You et al., 2018; Weiss et al., 2016)—not much because of the perceived physical safety of working alongside them, but rather because of the underlying concerns regarding whether work robotization would cause job losses (Weiss et al., 2016). As a matter of fact, a product not only can simultaneously evoke multiple emotions that may vary from person to person, but emotions are also, and foremost, the result of a deep intertwining between aesthetic qualities, perceived function, and brand reputation that altogether contribute forming a meaning into people's imaginary of a certain product application (Desmet, 2003). The aesthetic qualities of interaction and a product's configuration, however, can play a crucial role in shaping these imaginaries and in steering toward desirable emotions.

In the case of industrial robots, for instance, one can imagine that emphasizing only aspects of efficiency would render a robot fearful for workers, while explicitly communicating also how a robot, even when intended to operate autonomously, needs task planning, supervision, and maintenance–and thereafter also, needs humans–would make it more "pleasant."

But, *how do we turn a product, a robot especially, from a mere configuration of materials and parts into a communicative device able to convey specific meanings and elicit desired emotions?*

There is a rich and widely acknowledged body of design literature around product semantics and its dynamics. Colors, for instance, can be functionally used as accents to highlight specific components (Holtzschue, 2012) but also to elicit pleasant or unpleasant emotions (Valdez and Mehrabian, 1994). Shapes have also been found to have an effect on emotional responses, such as curve lines and organic forms being associated with soft and emotional products, while straight lines and sharp corners being associated with hard and rational responses (Hsiao & Chen, 2006). Even more, when leveraging a metaphorical language, i.e., the configuration of a product through the explicit reference to another thing presenting a relevant feature, form becomes functional and effective to communicate meanings and values, as well as to provide clues to users about product use (Cila et al., 2014). Relatedly, different materials have also been found to convey some meanings stronger than others, such as transparent and smooth materials being often used to convey the meaning of sexy, or hard and dark materials being used to convey a sense of professionalism (Karana et al., 2009).

Despite the extent and depth of knowledge on product semantics, the design of robotic artifacts, their appearance, and expressivity, are all but trivial. The choice of a color may easily raise questions of racialization (Bartneck et al., 2018), the resemblance of human-like or zoomorphic entities may encourage some to question whether the design is driven by actual needs or rather the wish for the capacity to replicate divine creation (Musa Giuliano, 2020). The peculiar history of robotics and its deep intertwining with the world of myths and legends (Longsdon, 1984), then, comes to the scene to challenge what we know about product design semantics. In their essence, however, robots distinguish themselves from other computational artifacts for their embedding in real-world environments, the capacity to respond to these, and also–and foremost–for their interface being inseparable from their functioning (Breazeal, 2004). In robotic artifacts, the tasks, interaction modalities, and expressive qualities are inseparable. This translates into a constant interplay between surface appearance and behavior, and *movement* more precisely (Hoffman and Ju, 2014). Movement, in fact, plays a prominent role in robot design as it is intrinsically present in the way these artifacts execute their tasks, afford interaction, and communicate a status.

Robot semantic, then, not only challenges traditional knowledge around product semantics through the display of peculiar morphologies, such as anthropomorphic or zoomorphic embodiments, but also adds movement as a key dimension to be considered when defining the artifact expressive qualities and channels for conveying meanings. This distinct expressive space also requires a broadening of design methodologies and perspectives. Traditional visual methods, such as sketching and 3D modeling, that have extensively been used to investigate the implications of specific product features (e.g., color or shape), keep been used also within the specific area of robot design. However, aspects of animacy and intentionality ask for novel ways to imagine not only how artifacts may look but also and foremost how these might behave. Thereafter, alternative perspectives on the shaping of robotic morphologies are increasingly being explored to challenge robot stereotypes and norms, and a growing number of performative methods are gaining momentum as practical ways to envision and experience early on how robots may appear and behave.

In the following sections of the chapter, we provide a collection of eight methods and perspectives illuminating how we can design robots' appearance and expressivity, accounting for their uniquely situated and dynamic nature.

In Box 2.1, Kwak and Kang suggest the *Robotization Design of Everyday Things* as a powerful approach to break free from the stereotypical imagination of robots as superhuman entities. The authors underscore the need of observing the patterns of people's everyday interactions with products as a way to identify specific functions worth of robotization. Adding selected robotic capabilities to everyday things with a specific function in mind, they argue, would help develop appearances that are better aligned with the functionalities and role of the robot, ultimately preventing dissatisfaction and frustration. To illustrate how the approach can translate into robotic artifacts, the authors present *PopupBot* a robotic pop-up space with diverse furniture modules embedded within its walls, which leverages an origami-based structure to open and close, adapting the space based on user intention.

In Box 2.2, Mandel and Ju present *Ready made Prototyping*. Similar to the previous, this method also seeks to break free from stereotypical ideas of robots and to build bridges with the world of everyday things. Yet, by emphasizing the mesh-ups between everyday objects and robotic components as the identity of the robots themselves, this approach brings a distinctively material perspective into the design of robotic artifacts. Robot appearance and expressivity become also the manifestation of a critical standpoint toward technology production. As the authors argue, when designing with ready-made objects, designers are invited to contemplate the forms, meanings, and affordances of existing things, and to clearly isolate the effects of the novel robot motion, sensing, or interaction, added through design. To illustrate the method, the authors present *HoverBot*, a robot platform based on recycled hoverboards that enabled the team to animate a series of existing products, such as chairs and trash barrels.

In Box 2.3, Baraka articulates the importance of designing *Multi-Modal Expressivity*. As we discussed, in fact, differently from traditional products, robots are artifacts uniquely characterized by a complex intertwining of expressive and functional elements. As the author explains, robot-to-human communication is made of a rich combination of signals, and these can be designed to encompass a variety of expressive modalities, many of which rely on non-verbal communication, such as gaze, gestures, prosody, posture, facial expressions, body movement, expressive lights, or sounds. To illustrate the importance of this approach, such as for communicating hidden robot states, the author reports on a project in which multi-modal expressivity was used for *robot-assisted therapy*. In this project, the use of multi-modal expressivity enabled the research team to engage children with autism in an inclusive manner by providing them a variety of expressive channels they could use to interpret the robot behaviors.

In Box 2.4, Zuckerman and Erel present *Minimalism* as a powerful perspective that breaks with robot design traditions. Predominantly characterized by a process of geometric abstraction, minimalism can help shape robots–their appearance, mechanism, and movement–in a reductive yet expressively rich way. As the authors discuss, instead of resembling familiar things or living beings, a minimalist appearance encourages users to build a personal interpretation of the robot, to make sense of it without the influence of stereotypical robot iconographies. Minimalism also encourage the designers to consider what is the minimum viable number of elements necessary to achieve a communicative result, whether it is in terms of motors in the mechanism or the amount and amplitude of the robot gestures.

To illustrate this perspective, the authors describe the project The Greeting Machine, an abstract non-humanoid robotic object intended to elicit either positive or negative reactions in the context of opening encounters. The project is inspired by social gestures commonly performed by people, like a gaze or head nod, but through the perspective of minimalism, these are translated into an elegant artifact presenting an embodiment made of only two white spheres which, through subtle movements, powerfully communicate in a way that resembles human non-verbal communication.

In Box 2.5, Ribeiro discusses *Animation* as a movement-centered method for the design of robot expressivity. While traditionally associated mostly with the world of 2D communication, such as cartoons and videogames, animation has actually proven of crucial importance in robot design too. As the author argues, the value of animation is to turn mere motion into a soul for the robot, which becomes a character in a story. Mostly leveraging non-verbal behaviors, animation principles allow to develop robots that do not need instructions to be understood, but rather promote mutual attention and implicit guidance. To illustrate how to put animation principles into robot design practice, the author presents the project *AvantSatie!* in which an autonomous Adelino robot play along the user in a pervasive game about music discovery.

In Box 2.6, Gallo and colleagues discuss *Bodystorming* as a valuable method for developing and evaluating robot design concepts. By leveraging role-play and physical enactment, bodystorming allows designers as well as research participants to enact robots that have not yet been implemented, enabling to explore a range of different user experiences involving interactions with robots, early on in the design process. As the authors argue, bodystorming is particularly useful for supporting interdisciplinary work, as it allows to lower the technical and conceptual barriers that may exist between professionals with different expertise and background. To illustrate how bodystorming can be applied in human–robot interaction (HRI) research, the authors report on a process carried out at the *Naver Labs headquarters*, where delivery robots are being developed as a service for the employees working in the building. Using bodystorming, the team enacted how the robot would operate, allowing to anticipate how people would respond and to identify practical interaction challenges.

In Box 2.7, Bleeker and Rozendaal present *Dramaturgy for Devices*, a theater-informed approach to designing the behavior of, and interaction with, robots and other intelligent artifacts. Similar to bodystorming and animation, this approach shifts the focus from imitation and representation as the basis for developing behavior and interaction toward performativity. It specifically invites to reflect on the meaning of behaviors and the sense of identity or character that emerges in the performance. In describing the method, the authors show how concepts from the theater, like for example, mise-en-scene, presence, and address, can be used to open the designer's eye to how situations afford (inter) actions, and (inter)actions afford interpretations. To illustrate this, the authors also describe a project in which they collaborated with theater professionals to enact situated encounters between humans and robots in a supermarket setting. Through this, they illustrate how dramaturgy can help understand how service robots might function as part of a social setting with customers.

In Box 2.8, Lee and colleagues describe *Worldbuilding*, a design approach in which a coherent and cohesive imaginary world that encompasses a multitude of artifacts, interactive elements, and contextual factors is constructed alongside the robot. The scope of this approach extends beyond the development of a single storyline. It rather offers context for design and informs about what elements need to be included in a project. As the authors argue, worldbuilding helps designers anticipate how people might interact with a robot, but also helps creating believability in the crafted interactions. As such, worldbuilding is a useful approach for both designers and people engaged in the crafted interactions. To illustrate how the approach translate into an actual robot design, the authors discuss their experience with designing *robot unboxing experiences*, in which they learned how children form mental models of the robot even before their direct interaction, starting from the moment they lay eyes on the delivery box, and continue forming while they unpack. Communication materials, packaging, and all possible kinds of artifacts related to a robot, then, contribute shaping the perception and opinions that people build about it, providing also the reference for interpreting expressive features.

BOX 2.1 THE ROBOTIZATION DESIGN OF EVERYDAY THINGS

Sonya S. Kwak and Dahyun Kang

Consumers often assess a product's category and form expectations regarding its functions based on appearance. When the product's actual functionality aligns with the expectations, the result is consumer satisfaction, but when it falls short, the results is disappointment. A robotic thing may fulfill consumer expectations generated by its appearance better than a human-like robot since its functionalities may not align with the expectations. Vacuum cleaning robots, resembling traditional vacuum cleaners, are categorized accordingly, with consumers expecting similar functions. Further, the integration of automation and robotic technologies often exceeds the expectations, promoting consumer acceptance. However, limited consumer acceptance of human-like robots can be attributed to two reasons. Firstly, categorizing these robots as living organisms leads to misaligned expectations with their actual functions, causing dissatisfaction. Secondly, difficulties arise when fitting human-like robots into existing product categories due to their human-like appearance. The expectation of versatile functions associated with humans complicates categorization, leading to dissatisfaction and hindered adoption.

In this vein, we propose **robotization of everyday things** as a design approach that could enhance consumer adoption of robots. In this design approach, designers can explore where robotization is required by observing users' way of living, including their usage patterns of everyday things; select the targeted everyday thing to robotize; choose the robotic technologies to apply for robotization; transform the everyday thing into a robotic thing by applying selected robotic technologies. A robotic thing that excels in categorization implies that its design is tailored to fulfill specific functions. However, such a specialized robotic thing may have limitations in comprehending user-contextual situations and offering multifunctionality in accordance with those situations.

To address these limitations, two design strategies may come in handy: on the one hand, **modularization** of forms and functions can let the robot adapt and transform to different forms and functions based on contextual situations, to meet user needs. On the other hand, designing robotic things for reciprocal **collaboration** can provide multifunctionality. By working together, robotic things can offer a wide range of capabilities and cater to diverse user requirements.

POPUPBOT

We designed PopupBot (Kwak et al., 2022) using the two strategies–modularization and collaboration–for **robotizing everyday things**. Existing non-transformable furniture occupies significant space when not in use, leading to wastage. PopupBot's furniture, equipped with robotic technology, can perceive environmental and user information, recognize situational contexts, and transform itself to efficiently utilize limited space.

Modularization

PopupBot is a robotic pop-up space with diverse furniture modules embedded within its walls. Through an origami-based design, these modules can fold and

FIGURE 2.1 PopupBot.

unfold like a pop-up book, adapting to different sizes and shapes to provide appropriate space based on user intention. The modularization of PopupBot can be categorized into three types: form-oriented, function-oriented, and form- and function-oriented. In form-oriented modularization, furniture modules adapt their form to support different situations, such as accommodating varying numbers of users or user body sizes. For example, PopupBot adjusts chair and desk length to accommodate single or multiple users, while the shelf expands or contracts to accommodate objects based on their number and size. In function-oriented modularization, furniture modules with the same form can serve different purposes by converting their function. For instance, if many people are invited to the PopupBot and require seating, the chair module alone may not be sufficient to accommodate everyone. In such cases, the PopupBot understands the situation, identifies which furniture module can support the chair module, and commands the side table to convert its function into a chair. In form- and function-oriented modularization, PopupBot changes its function by changing its form based on the user's intention for space utilization. For example, after a study session, the desk module can transform into a shelf for storing books and stationery by contracting its length, while the sofa module seamlessly converts into a bed by extending its length when the user intends to sleep.

Collaboration

To provide a diverse range of services previously performed by humanoids, robotic things need to collaborate with each other.

PopupBot facilitates collaboration among its modules, combining their capabilities to offer extensive services. This collaboration creates a dynamic space within PopupBot that seamlessly adapts to the user's intentions and requirements.

When a user awakens in PopupBot and expresses the desire to engage in study-related activities, the furniture and lighting components, comprising the desk, chair, bed, and associated illumination sources, synchronize their functionalities to transform a relaxation space into a productive learning space. This orchestration involves the simultaneous pop-up action of the desk and chair modules, folding of the bed unit, deactivation of the sleep light, and activation of the work light. Through this coordinated collaboration among the various modules of PopupBot, users can effectively utilize a singular space for multiple purposes, tailored to their specific intentions and needs.

BOX 2.2 READY MADE PROTOTYPING

Ilan Mandel and Wendy Ju

Ready made prototyping develops robot systems by adding sensing, automation, and feedback capabilities to existing products and infrastructure. This method allows designers to prototype how people might–or might not–interact with an automated world, rapidly and inexpensively.

In the art world, "**readymades**" are mass-produced objects transformed into art through the artists' selection and contextualization (Goldsmith, 1983). In HRI, the transformation of object to robot is more involved, but the emphasis is on using standard and inexpensive components to give the object the ability to move, or perceive people around them. This method is particularly apt for prototyping interactions with everyday objects which are being enhanced with robotic capabilities.

The use of existing objects helps **isolate the reaction** people are having to novel robot motion, sensing, or interaction, rather than the object or function. Examples of ready made prototyping in HRI include: Micbot, a microphone with two degrees of freedom, designed to increase group engagement (Tennent et al., 2019); Toasterbot, which expressively ejects toast as interaction and feedback (Ye et al., 2023); the Haunted Desk, which changes height when sedentary users should stand (Kim et al., 2021). Other examples include dinnerware, lamps, drawers, footstools, and chairs. Designing with ready made objects simultaneously affects the perspective of the designer. Rather than producing novel forms and purposes de novo, designers must contemplate the forms, meanings, and affordances of existing objects.

In doing so, they reason about existing patterns of interaction with everyday objects prior to augmentation and prototyping. This **on-going reflection** of the world enables HRI designers who are exploring new robot forms to quickly prototype forms from existing products. This method is often economical, because mass-produced objects are typically more affordable than low-volume quantities of the constituent materials, and are often available for low or no cost as waste. Ready made prototyping can thus also be more ecologically sustainable, by reusing existing products.

FROM HOVERBOARDS TO MOBILE ROBOTS

An example of **ready made prototyping** is our HoverBot Robot platform, which is based on recycled hoverboards. The hub motors, chassis, and battery of a hoverboard are augmented by a motor driver, and a Raspberry Pi

FIGURE 2.2 Trash barrel robot.

to provide a low-cost and quick-to-assemble mobile robot base able to move 80 pounds at 10 kilometers per hour. We use this platform to animate our Trash Barrel Robot (Bu et al., 2023) and Chairbot (Zamfirescu-Pereira et al., 2021).

We use this platform to animate our Trash Barrel Robot (Bu et al., 2023) and Chairbot (Zamfirescu-Pereira et al., 2021). We have also offered this as a platform to students in our mobile HRI design classes, and they have used it to make a range of robots, including a mobile umbrellabot, an autonomous medical crash cart, and a sample-tray robot.

Prior to developing HoverBot, we did what many roboticists do, which is use a modified vacuum cleaning robot as a platform. Turtlebot and iRobot's Create platform have long been used in both education and research contexts for prototyping indoor mobile robotics. The breadth of available documentation and tutorials makes them easy to use for a variety of robotics projects. Our previous version of the Trash Barrel Robot used a modified Neato vacuum robot, which had the benefit of being cheaper than the Create and came with a built in LiDAR. The need for the new motor platform stemmed from the fact that the vacuum robots did not move fast enough and/or with enough torque for our applications with trashbots, chairbots, and other robot furniture. We began looking toward using direct-drive Brushless DC (BLDC) motors, which have greater torque. For the cost of two motors and shipping from Aliexpress, which would take weeks to arrive, we could instead purchase a new hoverboard that used the identical motors, same day from local stores. The current cost for a new Swagtron T500 hoverboard is $99USD, whereas it would cost us $240USD to purchase the individual parts, even before accounting for shipping, taxes, and fees. Even better, we could often get these hoverboards for free, as many parents were happy to get rid of their children's used hoverboards for little to no money.

The initial effort to control the hoverboard took several iterations; one primary challenge lay in controlling the hub motors efficiently.

While we could flash custom code onto the original hoverboard control boards, this required tinkering across different brands and models. Using an open-source BLDC motor driver (ODrive v3.6), we could rapidly bring up new HoverBot platforms. We compensate for the heterogeneity among hoverboards from different manufacturers by doing a series of calibrations on the wheels. We have shared our instructions online at https://github.com/FAR-Lab/mobilehri2023 and verified that others were able to successfully follow these to make their own teleoperated and autonomous mobile robots. The HoverBot platform provides designers and researchers with the dynamic range to explore a wider variety of movements and interaction patterns. It turns out that the appropriate movement depends a lot on the objects placed atop the robot, and the context, as well as the users. **By making the platform easier and cheaper to build, the prototyping of the movement and interactions in-situ becomes easier to explore**.

BOX 2.3 MULTI-MODAL EXPRESSIVITY

Kim Baraka

Multi-modal expressivity is the broad idea that robot-to-human communication can be designed through a rich combination of signals pertaining to different expressive modalities. These modalities can be categorized into verbal (the use of speech) and non-verbal (the use of any other signal that can carry meaning). Non-verbal modalities of expressions have been an important focus of recent HRI research due to the important role they play in human-human and human-animal communication. They include, but are not limited to, gaze, gestures, prosody, posture, facial expressions, body movement, expressive lights (Baraka and Veloso, 2018) or sounds. A robot that combines multiple modalities could be perceived as more social, often resulting in more natural and seamless interactions with users.

While multi-modal robot behaviors may be hand-designed by experts from animation (Ribeiro and Paiva, 2012), dance, or domain-specific experts, recent work in social AI has looked at automating this process through generative multi-modal AI models trained on human data (Kucherenko et al., 2020) which are more flexible but not as reliable. For robots with a primarily functional role (e.g., manipulators, mobile robots), multi-modal expressivity is typically used to express hidden robot states that can **enhance human–robot collaboration**. For robots with a primarily social role (e.g., educational robots), multi-modal expression is typically used to **enhance a robot's perceived lifelikeness**, through evoking certain user perceptions or responses, often of an affective nature. In practice, there may be good reasons to avoid evoking user perceptions that do not align with the actual functioning of a robot (e.g., leading users from vulnerable groups to believe that a robot has actual feelings). Additionally, multiple overlapping communication channels may overwhelm or distract some users (e.g., users with sensory overload or hyperfocus).

Finally, note that some modalities are more costly than others and may require specialized hardware. For these reasons, multi-modal expressivity should always be analyzed and designed in context.

FIGURE 2.3 Robot-assisted therapy setup.

ADJUSTABLE MULTI-MODAL EXPRESSIVITY IN ROBOT-ASSISTED THERAPY

This project looks at the role of **multi-modality** in a therapy context. It was chosen to highlight the fact that having more expressive modalities does not necessarily imply better interaction. The project aims at targeting children with autism with the right modalities to allow them to pick up on non-verbal behavior of a robot. Children with autism often face difficulties decoding non-verbal behavior in humans.

As such, robots offer a simplified model of social interaction that can help them get exposed to such behavior in a controlled, simplified, and repeatable way, a stepping stone toward navigating human-human interactions more easily.

Central to the research is the idea of "just-right challenge" in a therapeutic context. This means that the robot must choose communicative actions that are neither too easy (at the risk of preventing learning) nor too difficult (at the risk of causing disengagement). In the tasks we designed, the more modalities the robot used to interact with the child, the easier the task is, since it becomes easier for the child to decode the communicative actions of the robot. As such, we investigate **what level of multi-modality should be used** for a particular child with a known level of autism-related impairment.

The modalities we consider are: speech, gaze, gestures, sound, and external visual stimuli. We furthermore look into how to adjust this number of modalities adaptively if the child is not able to understand the robot's behavior. The setup used for our robot-assisted intervention, shown in the figure, is an interactive storytelling scenario with screens showing video excerpts related to the story. Two interactive tasks are considered, in which the robot automatically adjusts its multi-modal behavior based on the levels shown in the table. The first one is a joint attention task aimed at directing the child's attention to one of the two screens, and the second one is a name-calling task, aimed at calling the child's attention back to the robot.

Our results highlight that different strategies for adjusting multi-modal behavior can lead to different interaction outcomes.

On the one hand, personalized and adaptive strategy developed in conjunction with therapists has the potential to promote learning. On the other hand, a random strategy is shown to be more effective in terms of immediate task performance. The key takeaway of this project in relation to expressive multi-modality is that **human–robot communication is a complex process** that can be shaped by appropriately designing for multi-modal expressivity according to interaction goals. Furthermore, this design process can go beyond simply authoring behaviors for each expressive modality, into having a robot intelligently reason about when and how to use these modalities in relation to context.

BOX 2.4 MINIMALISM

Oren Zuckerman and Hadas Erel

Minimalism is a principle with multiple interpretations in design and architecture, defined as "Less is More" by the Architect Ludwig Mies van der Rohe, as "Doing more with less" by the designer Buckminster Fuller, and is generally described as "reducing a subject to its necessary elements." Minimalism was inspired by Japanese traditional design, De Stijl artists, and the geometric abstractions associated with the Bauhaus movement. In the context of HRI, minimalism manifests in three aspects related to robots' design: the appearance, the mechanism and morphology, and the movement. The **robot's appearance design** involves critical decisions about the robot's form, should it resemble a human, an animal, a familiar object, or a non-familiar object. The minimalism principle in this case translates to a "minimal metaphor" experience, i.e., designing a form that users do not immediately associate with something familiar, influencing the user's association as minimally as possible, and hopefully leading to a deeper more intimate experience, as the user creates her own meaning with the absence of a leading metaphor. In "Less is More" terms, "less metaphor" will lead to "more experience." In our view, a non-familiar non-humanoid form would serve that goal in the best way, hence designing a non-humanoid form that has no immediate association, perceived as an abstract object at a first glance. Any design would impact perception and association, our approach is to lead toward associations of abstract geometric shapes, minimizing our bias on users' interpretations, and giving them the freedom to come up with their own associations and metaphors. The **robot's mechanism and morphology design** includes the number of motors and the mechanical design, together forming the robot's Degrees of Freedom (DoF) and range of possible movement. The goal is to generate subtle and gentle gestures, in our opinion a minimal mechanism design is more elegant, and has greater chances to result in fluent movement, "doing more with less." Minimalism of mechanism and morphology design translates to a minimum number of motors, and a creative design of gears and mechanisms to generate the desired movement. The **robot's movement design** includes the gestures characteristics: range of movement, speed, and duration. Minimalism of movement design translates as "reducing a gesture to its necessary movement elements." With gestures designed according to psychological principles, the impact of very minimal movement can be immense, not only on perception, but also on awareness, behavior, emotions, and even fundamental psychological needs.

FIGURE 2.4 The greeting machine, prototypes, and mechanisms.

THE GREETING MACHINE

The Greeting Machine (Anderson-Bashan et al., 2018) is an abstract non-humanoid robotic object. The goal was to communicate positive and negative social cues in the context of opening encounters, inspired by social gestures commonly performed by people, like a slight gaze or head nod, or a subtle change in body orientation. In our design process, we followed our "**minimal metaphor minimal movement**" approach, striving for a unique design that has no direct metaphor of human, animal, or familiar object.

In our design process we explored **basic geometric shapes** using low-fidelity prototypes and simple animation studies, and gradually leaned toward curvilinear forms. The final design included a dome-shaped object as a static base, with a small ball that we envisioned to roll all across the larger dome's surface. We used white color to minimize undesired associations of specific colors, and 3D printed plastics (PLA/ABS) as it was extremely hard to form a custom empty sphere from wood. With regards to the mechanism design, we strived for a minimal number of motors that will somehow enable the ball to move all across the dome surface. In a long process of experimentation, our team reached an elegant 2-DoF design consisting of a base rotation and a tilt lever, supporting a polar coordinate mechanism by leveraging magnetic force at the edge of the lever to roll the small ball on the sphere's surface. Regarding the movement design, we strived for minimal gestures. Since there is **no direct mapping between a human's gesture of opening encounter and a non-humanoid abstract robot**, we invited several movement experts, including an animator, a puppeteer, and a choreographer, to show us what the movement of the small ball should be in various situations of opening encounters, using a low-fidelity prototype.

They discussed and debated various movement characteristics, including start and end positions, style and pace of the movement, straight vs. curved trajectories, and concluded with two main gestures: Approach (back-to-front) and Avoid (front-to-back). We implemented these gestures, and in a qualitative study with 26 participants, we were able to validate many of our design assumptions.

Regarding the "minimal metaphor" principle, participants indeed had no direct metaphor for it: "It was different.. it's just a thing", "It looks futuristic, like nothing from this real world or from this decade, like something from the distant future", "Its spherical form makes it kind of calm", "very clean, white, and circle, and a little ball."

Regarding the "minimal movement" principle, participants indeed reacted to the Approach and Avoid gestures in a very emotional way: "When it was turning and facing me then I thought it was really welcoming...", "I had the feeling he, it, is avoiding me, like it feels uncomfortable. That's why it wants to turn away."

Since then, in a series of studies with several non-humanoid robots we designed, we validated how impactful minimal gestures can be. For example, a robotic object designed as four slightly-moving pillars provided a sense of companionship for older adults (Zuckerman et al., 2020), and a lamp-like robot performing minimal gestures enhanced the sense of security and encouraged exploration of novel experiences (Manor et al., 2022).

BOX 2.5 ANIMATION

Tiago Ribeiro

Animation (Ribeiro and Paiva, 2012) isn't just about motion. It means to provide a **soul** to an embodiment. To tell a **story**, turning an embodiment into a character, expressing its identity, thoughts, and underlying motivations. All of this applies to social robots.

When developing a social robot, one must consider three parts: the animated embodiment along with its form and capabilities; the target audience, who they are, or expected to be; and where the task occurs (the stage). The task, or application, lies within the **intersection** of the three and should be design-centered into that space. In part, it requires thinking of the interaction as an act. With an introduction of who the robot is, what it can do, how the audience can interact with it. This is where it expresses its identity and underlying motivation and sets the expectations of the user. *(The story is all about managing expectations)*. Is it there to *entertain*? To *assist* in a task? To provide *emotional support*?

The *act* begins. A quick demonstration of its own capabilities may be used to introduce the task and how the two parts can interact. Demonstrate the robot's *acknowledgment* of the **users** and of its own **self**. With non-verbal robots, ensure there is a display in the task to provide guidance and *mutual attention*, instead of a human providing instructions. Do the best to *show*, not tell. Design the whole act with an artistic vision. Implement the task alongside **animators** and **designers**. Perform iterative **usability tests** to ensure users understand the activity. Use 3D game engines for interaction-previsualization, and their animation systems to drive the robot, blending between faithful animation playback, and goal-driven procedural motion such as IK, pick-and-place, or locomotion.

Interaction Design for HRI is a cousin of Game Design. Taking advantage of **existing tools** is key to a successful design.

AVANTSATIE!

AvantSatie! is a pervasive game where players must discover the musical score of a piece using a floor piano. An autonomous Adelino robot helps them by performing expressive gazing and animations. The robot uses the ERIK technique to track the player's face, and to gaze at specific piano keys, while shaping its posture, in order to convey hints to the player.

FIGURE 2.5 Various screenshots of the projected screen of the *AvantSatie* game.

By playing the piano keys, following instructions, observing the robot, and understanding its hints, participants can minimize the mistakes performed through the trial-and-error nature of the discovery task. We designed AvantSatie iteratively until the pilot players were able to understand the game with no human introduction or assistance, relying solely on screen-instructions and Adelino's non-verbal behavior. Aiming to study if ERIK could provide

Adelino with the ability to communicate non-verbal hints to a user about what action to take next during a particular task, while also gaze-tracking the user and **conveying the illusion of life**, we ran an experiment comparing three versions of AvantSatie. In versions C-ERIK and C-EBPS, Adelino conveyed a *"warm-cold"* heuristic of each player's guess by shifting between *Neutral, Warm* and *Cold* expressive postures, while keeping the gaze-tracking behavior toward the player. C-ERIK used ERIK to perform the gaze-tracking with an expressive posture. C-EBPS used a non-IK posture synthesis technique that interpolates a large number of manually authored postures for each expression through a range of gaze directions. C-Control differs from the other two conditions only by not performing any posture-expressive behavior at all; the robot still uses ERIK for gaze-tracking but remains in the Neutral posture throughout the whole game.

Study participants were given no indication of the robot's behavior and were solely informed that they should follow the instructions on the screen and observe the robot when performing each guess. The robot was already active when they entered the room and would start face-tracking them once they stepped into the Kinect's view, to convey the robot as an autonomous entity and immerse them into the game. Subjective measures were collected based on the following questionnaires:

PMU and CP from Networked Minds; RoSAS, *Perceived Adaptability* from the *Almere model*; plus, the custom-designed questionnaires *Robot's Performance and Usability, Robot's Intention and Motivation*, and *Animation Illusion of Life*. We additionally collected objective data exclusively during the *Guessing* phases, consisting of Time Spent Guessing, WrongHot (# incorrect guesses assessed as Hot), WrongCold (incorrect guesses assessed as Cold), and WrongTotal (WrongHot + WrongCold).

The results allowed to conclude that the participants noted the robot's intention-directed postural behavior when using ERIK, and were able to intuitively understand it; the results when using ERIK were similar to those of a manually tailored (and laborious) alternative technique; players' understanding and playability of the game was not impacted by study conditions, revealing an adequate interaction design; ERIK succeeded in making the robot convey the intended expressivity, and the illusion of life despite the shakiness of the low-fidelity craft robot.

BOX 2.6 BODYSTORMING

Danilo Gallo, Maria Antonietta Grasso, Kahyeon Kim, and Sure Bak

Bodystorming (Schleicher et al., 2010) is a design method that involves role-playing and physical enactment of scenarios to explore and evaluate design concepts in a more realistic and tangible way before technical implementation.

The origins of bodystorming can be traced back to the field of Human–Computer Interaction (HCI) in the early 1990s. The approach was developed as a way to address the limitations of more traditional design methods, which often relied on abstract models rather than real-world interactions. It has mainly been used in two ways. The first one, a simpler one, consists in **designing while immersed in the final environment of use**, to consider contextual elements affecting the interaction. The second one, includes the presence of actors and props to creatively simulate various experiences of using a technology. The second one is naturally more creative and exploratory in nature and has been successfully adopted in HRI because of the advantages it offers in terms of interdisciplinarity and creativity.

In the context of HRI, bodystorming allows participants to **enact robots that have not yet been implemented**. This makes it possible to explore a range of different user experiences involving interactions with robots, such as handing objects or co-navigating in shared spaces. Bodystorming is particularly useful for supporting interdisciplinarity, as it allows designers and developers to explore together without the lengthy and costly development of the robot prototypes. By physically enacting scenarios and interactions, designers can directly leverage their natural intuition about the requirements of the user experience in the envisioned scenarios. At the same time, the involvement of participants with technical profiles allows to identify potential issues or areas for technology improvement.

Overall, bodystorming is a powerful tool for designers and developers working in the field of HRI. It enables them to explore and evaluate robot design concepts in a more realistic and tangible way before investing significant resources in technical development.

SCENARIOS OF INDOOR DELIVERY ROBOTS

Naver Labs has developed and deployed a robotic platform at Naver's headquarters, a high-rise robot-friendly office building, to carry out delivery tasks for employees, including delivering parcels, coffee, and food to offices and meeting rooms. Our teams have used **bodystorming**, both to inform the design of the robot service as well as to research fundamental HRI questions. We present two use cases illustrating the flexibility of the bodystorming method.

FIGURE 2.6 The bodystorming session took place in the Parcel Delivery Center. This was the real space where, once deployed, the robots would be loaded with parcels for delivery by warehouse employees. Each person carried a sign indicating their role in the session, i.e., robot (including a number) or control system. Those enacting robots received a quick orientation on the the rules of movement (e.g., speed) before starting the session.

The first use case (Gallo et al., 2022) focused on investigating socially acceptable robot behaviors for shared elevators and explored fundamental HRI questions about different approaches to designing them, i.e., imitating humans (human-like) or differentiating from them (machine-like). The team video recorded a two-hour bodystorming session with four researchers from design and technical disciplines involved in the project. One person at a time performed the role of a robot and lead researchers asked participants to act out different elevator use situations based on previous ethnographic observations. Each participant started by enacting a robot displaying human-like behaviors. Then, alternative machine-like behaviors were explored. In the latter, participants considered robot capabilities and limitations, e.g., access to sensor information and physical characteristics of the robot designed by our company, i.e., non-humanoid, without arms, and including displays, light, movement, and sound to communicate its intentions. After each situation, all participants shared their impressions and analyzed the actions performed with the researchers. The recorded session was later analyzed by the lead researchers, and data was clustered to form common themes such as position, distance, movements, priority, interaction, preferences, risks, and feedback modes.

Bodystorming allowed to **identify the strengths and limitations** of both approaches and the actions of the participants provided guidelines for the elements of human behavior and machine behavior that the robot should follow in each situation. The second use case informed the design of the multi-robot control system, ARC Brain, which manages the fleet of robots in the headquarters building. The goal of the bodystorming session was to identify potential routing problems before its development.

There are five positions in which delivery robots wait to load parcels in the warehouse. When a robot departs with a parcel, another robot, which is waiting on queueing line, would take the empty position. The route from queueing line to departure gate is one-way. So, if one robot fails, everything could be stopped. The session focused on identifying bottlenecks or areas where there was a high risk of collision. The session lasted three hours and involved 12 participants, including developers, project managers, and designers. Ten participants enacted the robots. The two remaining ones were ARC Brain designers and took on the role of the system, giving instructions to enacted robots. Lead researchers specified several robot control scenarios for which is complex to **anticipate problems**.

Enacted robots moved following the (enacted) system's guide. When one scenario was completed, there was a short session to share thoughts. After the session, a survey and a discussion took place. The bodystorming activity not only enabled to identify potential problems, but also contributed to build consensus among stakeholders on the directions to take for preventing those issues.

BOX 2.7 DRAMATURGY FOR DEVICES

Maaike Bleeker and Marco C. Rozendaal

Dramaturgy for devices (Bleeker and Rozendaal, 2021) is a theater-informed approach to designing the behavior of, and interaction with, robots and other intelligent artifacts. This approach offers a *perspective* from which to analyze robot behavior and HRI, *conceptual tools* to describe and interpret behavior and interaction, and *methods and skills* to develop them. These are brought together in the Dramaturgy for Devices Toolbox that is currently being developed.

Dramaturgy for Devices as *perspective* builds on proposals from the fields of sociology, anthropology, gender studies, and performance studies to conceive of everyday interaction as a kind of performance. This approach **shifts focus from imitation and representation** as basis for developing behavior and interaction **toward performativity**: how do the meaning of behavior and a sense of identity or character emerge in the doing? Performativity, thus, is not a matter of what robots or other intelligent objects do per se (their performance), but describes what this doing brings about within the given situation. From this perspective, and following Barad's (2007) understanding of posthumanist performativity, we might say that it is within a particular network of relationships between humans and things that robots and other intelligent technologies gain the agency to (inter)act in a meaningful way.

Dramaturgy for Devices shows how *concepts* from the theater, like for example mise-en-scene, presence, and address, can be used to open the designer's eye to how situations afford (inter) actions, and (inter)actions afford interpretations. This includes, but is not limited to, how what may appear as the "character" and the intentions of a smart technology, are actually the effect of what it does and how, and how what it does can be interpreted within the given situation. Dramaturgy for Devices also shows how *methods and skills* from the theater, including improvisation, puppeteering, choreography, scenography, acting, and scriptwriting, can inform new performative approaches to designing behavior and interaction as situated and relational phenomena.

SPECULATIVE ENCOUNTERS WITH SUPERMARKET ROBOTS

Using our **Dramaturgy for Devices** approach, we collaborated with theater professionals to enact situated encounters (Gemeinboeck 2021) between humans and robots in a supermarket setting. The workshop was part of a larger research program that works toward the further development of PAL Tiago robots.

FIGURE 2.7 The mixed-reality setup with an actor using virtual reality in front, and their robot avatar in the virtual supermarket in the back.

Whereas other parts of this research program focused on technological developments, the focus of our contribution was on understanding how these service robots could function as part of a social setting with customers. To this end, we developed a mixed-reality setup in that combines VR with real-life interaction, and in which actors performed embodied explorations of human–robot encounters while a puppeteer controlled the robot. The actor's training provided them with the skills necessary to perform life-like interactions and to repeat these with variations. Furthermore, actors are trained not only to convincingly perform ways of behaving but also to invent behavior that feels right in a particular situation and in interaction with others. They are experienced in making their empathic and emotional responses part of their "reading" of the situation, including their own position and possibilities for action within it. They were thus capable of inventing different ways of responding and interacting and of imagining how they might respond in different ways and in different situations. This is not the same as an uninformed test-person responding spontaneously to an unknown situation, but this was also not the aim of our investigation. Rather, our aim was to map possible responses and understanding how these are brought about.

The **creative as well as analytical skills of the actors and the puppeteer** allowed us to experiment with novel robot behaviors and speculate on how these may afford HRI. We were interested in particular in the methodological potential and implications of our setup for new approaches to the development of HRI that are not based on making robots execute pregiven scenario's but instead build on possibilities for emergent behavior that is situated (i.e., specific to the situation and the robot's morphology).

In this context, the actor's skills proved to be most useful for investigating how situated encounters with robots may trigger responses and bring about interaction. Furthermore, we worked with actors who had worked in drama-based training situations in which they had learned to verbalize detailed observations and reflections about the interactions that they were part of. They were thus capable of using insights and concepts from the theater to contribute to a collective analysis of what happened in the situated encounters.

The puppeteer brought in a different set of theater skills. Unlike the actors, he was not shaping his own performance but animating another entity from the outside. As he put it: "in puppeteering I am supportive to what I need to control. I am not the actor, rather, I support a story by *bringing something to life*." His experience as puppeteer informed an approach to animating the robot from a creative exploration of the morphology of the robot and its possibilities for movement, rather than from imposing human-like motions on it. This proved most useful for understanding how robotic bodies and their behavior can **generate new modes of interacting** that do not follow existing human models but instead follow from their own morphology and modes of expression.

BOX 2.8 WORLDBUILDING

Christine P. Lee, Bengisu Cagiltay, and Bilge Mutlu

Worldbuilding refers to the process of constructing a coherent and cohesive imaginary world that encompasses a multitude of artifacts, interactive elements, and contextual factors. Its scope extends beyond the development of a single storyline, as these worlds are built around multiple interconnected components with deliberate rhetorical intentions (Coulton et al., 2017). Worldbuilding offers a platform to explore and engage with the rhetorical features embedded within the world, fostering freely navigated interactions that are not confined to a prescribed path (Coulton et al., 2017, Zaidi, 2019). The creative endeavor of worldbuilding involves the crafting and sculpting of diverse artifacts and prototypes, such as narratives, characters, social, and cultural context, and plots. Through the construction of these details, worldbuilding consists of creating new artifacts and prototypes that serve as blueprints throughout an individual's interaction, which evokes a wide range of immersive experiences and distinct perspectives.

The constructed "worlds" can ultimately **offer context for design**, inform what elements must be included in the design, and help designers predict how people might interact with their designs. The coherent integration of the various components within worldbuilding plays a vital role in establishing believability (Mutlu, 2021). Given the imaginative nature of worldbuilding, the seamless interplay between all elements involved in the experience, including characters, narratives, events, and interactive dynamics, is essential for users to fully comprehend and engage with the constructed world. The level of **believability** significantly influences the immersive nature of the experience, hinging on factors such as how the user and features in the imaginary world interact or how the storyline evolves.

Worldbuilding as a design research method enables designers to ideate novel HRI for diverse user groups within a range of contexts (Simmons et al., 2011; Lee et al., 2022). This method enables solution development across different interaction stages, including before, during, and after. Worldbuilding might be a resource not only for supporting users' perceptions and experiences, but also for supporting HRI researchers' and designers' creativity in designing robot interactions. For example, researchers and designers can transfer principles from other creative domains, including performing arts, theater, film, or improv, to the practice of building the robot's world.

A SOCIAL ROBOT FOR CHILDREN

We explored **worldbuilding** through observing natural interactions and co-design, with a specific focus on creating a curated social robot unboxing experience tailored for children. We first examined children's inherent unboxing behaviors and encounters, extracting insights to construct a tailored unboxing experience aligned with their contextual requisites. This stage captured their

FIGURE 2.8 Children Interacting with the designed unboxing experience for a social robot.

interaction patterns, instinctive reactions, habits, and perceptions when engaging with the robot.

The exploration revealed that children formed mental models of the robot even before their direct interaction, starting from the moment they laid eyes on the robot's delivery box, and continued forming their mental models as they unpacked it. This highlighted the significant potential of worldbuilding to **enhance the robot experience by leveraging opportunities that extended beyond the actual robot interaction**. It also underscored that once a world or mental model was constructed, and children engaged with the robot during interaction, the impact persisted even after the interaction had concluded, further immersing the child within the constructed world. We then conducted co-design sessions with children to collaboratively shape the specific aspects of the world, encompassing artifacts, interactions, characters, narratives, and more. This process offered valuable insights, highlighting the necessity and specific design of components in the constructed world for the unboxing experience.

The final unboxing experience featured a "box" with a distinctive social presence and character, concurrently functioning as the robot's home. The narrative encompassed the box's career transformation from its retirement as a social robot to a butler, now entrusted with the role of facilitating the initial interactions between the child and the robot. Moreover, the design of artifacts, such as the design of the instruction manuals and the box, was designed to align with the constructed world for seamless integration. Furthermore, a four-phase interaction procedure, including prior interaction, packaging, first interaction, and first impression, was formulated with a diverse array of activities to enhance the quality of the initial interaction between children and social robots.

The case study exemplifies the imperative and significant potential of worldbuilding. Their approach involves a comprehensive **understanding of the contextual intricacies**, collaborative co-construction of the world's constituents with end users, and the creation of tangible artifacts to integrate into the constructed world. These procedural stages offer valuable insights for other designers, who may opt to integrate elements or the entirety of the observation, co-design, and prototyping processes into their own worldbuilding endeavors to provide a more immersive robot interaction experience. Further, the scope of worldbuilding extends from seemingly mundane routines, including the robot's delivery process to a child's doorstep, to the smallest technical aspects, such as the robot's on-off switch. Thus, worldbuilding holds the transformative potential to reshape users' interactive experiences and fundamentally **reshape their perceptions of the robot** at hand.

REFERENCES

Anderson-Bashan, L., Megidish, B., Erel, H., Wald, I., Hoffman, G., Zuckerman, O., & Grishko, A. (2018, August). The greeting machine: an abstract robotic object for opening encounters. In *2018 27th IEEE international symposium on robot and human interactive communication (RO-MAN)* (pp. 595–602). IEEE.

Barad, K. (2007). *Meeting the universe halfway: Quantum physics and the entanglement of matter and meaning*. Duke University Press.

Baraka, K., & Veloso, M. M. (2018). Mobile service robot state revealing through expressive lights: formalism, design, and evaluation. *International Journal of Social Robotics, 10*, 65–92.

Bartneck, C., Yogeeswaran, K., Ser, Q. M., Woodward, G., Sparrow, R., Wang, S., & Eyssel, F. (2018, February). Robots and racism. In *Proceedings of the 2018 ACM/IEEE international conference on human-robot interaction* (pp. 196–204).

Bleeker, M., & Rozendaal, M. C. (2021). Dramaturgy for devices: Theatre as perspective on the design of smart objects. In M. C. Rozendaal, B. Marenko, & W. Odom (Eds.), *Designing smart object in everyday life: Intelligences, agencies, ecologies* (pp. 43–56). Bloomsbury.

Breazeal, C. (2004). Social interactions in HRI: The robot view. *IEEE Transactions on Systems, Man, and Cybernetics, Part C (Applications and Reviews), 34*(2), 181–186.

Bu, F., Mandel, I., Lee, W. Y., & Ju, W. (2023, March). Trash barrel robots in the city. In *Companion of the 2023 ACM/IEEE international conference on human-robot interaction* (pp. 875–877).

Calo, C. J., Hunt-Bull, N., Lewis, L., & Metzler, T. (2011, August). Ethical implications of using the paro robot, with a focus on dementia patient care. In *Workshops at the twenty-fifth AAAI conference on artificial intelligence*.

Cila, N., Hekkert, P., & Visch, V. (2014). Source selection in product metaphor generation: The effects of salience and relatedness. *International Journal of Design*, *8*(1).

Coulton, P., Lindley, J., Sturdee, M., & Stead, M. (2017, March). Design fiction as world building. In *Proceedings of the 3nd biennial research through design conference* (pp. 1–16).

Crilly, N., Moultrie, J., & Clarkson, P. J. (2004). Seeing things: consumer response to the visual domain in product design. *Design Studies*, *25*(6), 547–577.

Demirbilek, O., & Sener, B. (2003). Product design, semantics and emotional response. *Ergonomics*, *46*(13–14), 1346–1360.

Desmet, P. (2003). A multi layered model of product emotions. *The Design Journal*, *6*(2), 4–13.

Gallo, D., Gonzalez-Jimenez, S., Grasso, M. A., Boulard, C., & Colombino, T. (2022, March). Exploring machine-like behaviors for socially acceptable robot navigation in elevators. In *2022 17th ACM/IEEE international conference on human-robot interaction (HRI)* (pp. 130–138). IEEE.

Gemeinboeck, P. (2021). The aesthetics of encounter: A relational-performative design approach to human-robot interaction. *Frontiers in Robotics and AI*, *7*, 577900. https://doi.org/10.3389/frobt.2020.577900

Goldsmith, S. (1983). The readymades of Marcel Duchamp: The ambiguities of an aesthetic revolution. *The Journal of Aesthetics and Art Criticism*, *42*(2), 197–208.

Hoffman, G., & Ju, W. (2014). Designing robots with movement in mind. *Journal of Human-Robot Interaction*, *3*(1), 91–122.

Holtzschue, L. (2012). *Understanding color: an introduction for designers*. John Wiley & Sons.

Hsiao, K. A., & Chen, L. L. (2006). Fundamental dimensions of affective responses to product shapes. *International Journal of Industrial Ergonomics*, *36*(6), 553–564.

Karana, E., Hekkert, P., & Kandachar, P. (2009). Meanings of materials through sensorial properties and manufacturing processes. *Materials & Design, 30*(7), 2778–2784.

Kim, L. H., Leon, A. A., Sankararaman, G., Jones, B. M., Saha, G., Spyropolous, A., … Paredes, P. E. (2021, March). The haunted desk: exploring non-volitional behavior change with everyday robotics. In *Companion of the 2021 ACM/IEEE international conference on human-robot interaction* (pp. 71–75).

Krippendorff, K., & Butter, R. (1984). Product semantics-exploring the symbolic qualities of form. *Departmental Papers (ASC)*, 40.

Kucherenko, T., Jonell, P., Van Waveren, S., Henter, G. E., Alexandersson, S., Leite, I., & Kjellström, H. (2020, October). Gesticulator: A framework for semantically-aware speech-driven gesture generation. In *Proceedings of the 2020 international conference on multimodal interaction* (pp. 242–250).

Kwak, S. S., Park, S., Kang, D., Lee, H., Yang, J. H., Lim, Y., & Song, K. (2022, March). PopupBot, a robotic pop-up space for children: Origami-based transformable robotic playhouse recognizing children's intention. In *2022 17th ACM/IEEE international conference on human-robot interaction (HRI)* (pp. 1196–1197).

Lee, C. P., Cagiltay, B., & Mutlu, B. (2022, April). The unboxing experience: Exploration and design of initial interactions between children and social robots. In *Proceedings of the 2022 CHI conference on human factors in computing systems* (pp. 1–14).

Longsdon, T. (1984). *The robot revolution.* Simon & Schuster Inc., New York.

Manor, A., Megidish, B., Todress, E., Mikulincer, M., & Erel, H. (2022, August). A non-humanoid robotic object for providing a sense of security. In *2022 31st IEEE international conference on robot and human interactive communication (RO-MAN)* (pp. 1520–1527). IEEE.

Monö, R. G., Knight, M., & Monö, R. (1997). *Design for product understanding: The aesthetics of design from a semiotic approach.* Liber.

Morris, P. H., Reddy, V., & Bunting, R. C. (1995). The survival of the cutest: who's responsible for the evolution of the teddy bear? *Animal Behaviour, 50*(6), 1697–1700.

Musa Giuliano, R. (2020). Echoes of myth and magic in the language of artificial intelligence. *AI & Society, 35*(4), 1009–1024.

Mutlu, B. (2021). The virtual and the physical: two frames of mind. *Iscience, 24*(2), 1–16.

Ribeiro, T., & Paiva, A. (2012, March). The illusion of robotic life: Principles and practices of animation for robots. In *Proceedings of the seventh annual ACM/IEEE international conference on human-robot interaction* (pp. 383–390).

Schleicher, D., Jones, P., & Kachur, O. (2010). Bodystorming as embodied designing. *Interactions, 17*(6), 47–51.

Simmons, R., Makatchev, M., Kirby, R., Lee, M. K., Fanaswala, I., Browning, B., … Sakr, M. (2011). Believable robot characters. *AI Magazine, 32*(4), 39–52.

Tennent, H., Shen, S., & Jung, M. (2019, March). Micbot: A peripheral robotic object to shape conversational dynamics and team performance. In *2019 14th ACM/IEEE international conference on human-robot interaction (HRI)* (pp. 133–142). IEEE.

Valdez, P., & Mehrabian, A. (1994). Effects of color on emotions. *Journal of Experimental Psychology: General, 123*(4), 394.

Weiss, A., Huber, A., Minichberger, J., & Ikeda, M. (2016). First application of robot teaching in an existing industry 4.0 environment: Does it really work? *Societies, 6*(3), 20.

Ye, M., Lee, R., Michalove, J., & Wong, J. (2023, March). Toaster bot: Designing for utility and enjoyability in the kitchen space. In *Companion of the 2023 ACM/IEEE international conference on human-robot interaction* (pp. 787–790).

You, S., Kim, J. H., Lee, S., Kamat, V., & Robert Jr, L. P. (2018). Enhancing perceived safety in human–robot collaborative construction using immersive virtual environments. *Automation in Construction, 96*, 161–170.

Zaidi, L. (2019). Worldbuilding in Science Fiction, Foresight and Design. *Journal of Futures Studies, 23*(4), (pp. 15–26).

Zamfirescu-Pereira, J. D., Sirkin, D., Goedicke, D., Lc, R., Friedman, N., Mandel, I., … Ju, W. (2021, March). Fake it to make it: Exploratory prototyping in HRI. In *Companion of the 2021 ACM/IEEE international conference on human-robot interaction* (pp. 19–28).

Zuckerman, O., Walker, D., Grishko, A., Moran, T., Levy, C., Lisak, B., … Erel, H. (2020, April). Companionship is not a function: The effect of a novel robotic object on healthy older adults' feelings of "Being-seen". In *Proceedings of the 2020 CHI conference on human factors in computing systems* (pp. 1–14).

3 Designing for Social Embeddedness

Mutually Shaping Robots and Society

Selma Šabanović
Indiana University, Bloomington, United States

The idea that the adoption and use of robots will affect our work, relationships, and other aspects of daily life is widely imagined in science fiction, commented on in scholarship, and accepted by the public. We are all aware of, even if we do not share, anxieties regarding the potential of robots to replace us in the workplace (Mindell, 2015), warnings that interacting with robots will leave us more socially isolated and lonely (Turkle, 2011), and hopes that wider access to robots may solve some of our biggest societal issues, from aging societies to the need for safer and more convenient transportation (Wright, 2023). These hopes and fears, however, often assume that social change will come directly from the mere availability of robots in our daily lives.

As robots have started being deployed and evaluated in everyday contexts of use, it has become clear that robots alone are not the drivers of social changes and that their effects and usefulness are intricately tied to the social contexts of their development and use. The social and cultural values and assumptions built into the design affordances of robots interact with the social and cultural aspects in use contexts, creating the possibility of unexpected emergent interactions and societal consequences. Mutlu and Forlizzi's (2008) ethnographic study of a delivery robot in a hospital setting showed that the type of patient being served in a hospital wing and the resulting intensity and emotional valence of the work being done there can be the deciding factor in whether nursing staff appreciate or abhor their robot helpers. Lee and her coauthors (2016), on the other hand, found that robot designs based on the assumption that aging is primarily a deficit in cognitive and physical function were often unpalatable to older adults, who preferred technologies that could scaffold the more positive and joyful aspects of aging. These and many other user evaluations of robots suggest that values embedded in robot design and social factors in the environment are central in defining ensuing human–robot interactions (HRI), user experiences, and resulting societal outcomes.

DOI: 10.1201/9781003371021-3
This chapter has been made available under a CC-BY-NC-ND license.

The interactions between the social factors that affect robot design and those found in their contexts of use can create a feedback loop that is positive; Wan-Ling Chang and I saw this dynamic among nursing staff who saw the benefits of using the seal-like Paro robot with older adults in an eldercare facility and were afterward able to come up with their own ways of beneficially using the robot (Chang and Šabanović, 2015). It can also lead to more negative experiences and outcomes, such as early adopters' tendency to over-trust autonomous vehicle capabilities, leading to a lack of attention and control in difficult driving contexts that have led to accidents and even deaths. Even positive direct interactions with robots can still lead to overall negative social outcomes, such as in the case of the German older adult living in a nursing home depicted in the 2007 documentary film "Mechanical Love". The movie depicts an older woman's successful use and connection with the seal-like robot Paro in her room, and then pans out to the broader picture showing the annoyance of other residents of the facility at the robot's presence, the older woman's resulting distancing from others, and concerns by the staff whether such a connection between an older adult and a robot is healthy.

These examples make clear that it is not sufficient to develop robots first and only then figure out what the resulting social perceptions and their effects might be. Taking the social factors that shape robot development and use into consideration in the design process requires approaches to design that incorporate an understanding of the "mutual shaping" feedback loop between society and robotic technologies in the design process (Šabanović, 2010) Traditionally, the focus of robot design has been on the complex body and infrastructure of the robot itself – its technical components, appearance, the behaviors that it needs to have to interact with its environment. Efforts at exploring what people might think of and how they will react to the robot's appearance and behaviors coming later. This technocentric design process, however, does not take into consideration that a robot is "socially embedded" in all phases of its design – from the initial design idea, to the implementation of that idea in a physical interactive form, and to the evaluations of the system and further adaptations that occur later on in the context of use. Designing a robot, therefore, requires that we think beyond the robot itself throughout the design process, to address how social context, norms, and values play a role in its design and later use. This means thinking through how different design features of the robot will affect and be affected by aspects of the social context of use, which can include the social dynamics between people, social and cultural norms that guide behaviors and their interpretation, and factors beyond the HRI that affect the priorities and needs of potential users.

Design is inherently a generative, future-looking endeavor, aiming to create new possibilities for interaction out of existing or even imagined materials. To design a new technology or interaction, we must envision and critically evaluate user experiences that are not only unfamiliar, but don't yet fully exist. How can designers, potential users, and other stakeholders that will be affected by technology imagine and put to the test these various possible futures? Though we know that robotics and society will mutually shape each other, how do we explore what form that interaction will take with technologies that we do not yet have, so we can evaluate their potential benefits and pitfalls? Several of the methods outlined in this section suggest ways in which designers can present potential users with the possibilities of new technologies and experiences to get their feedback and provide insights that can guide further design.

"Breaching experiments" (Box 3.6) can provide a way of bringing social and cultural norms to the fore through moments of rupture so that they can be more consciously incorporated into the design. Bødker suggests that having a robot in an everyday social context is in itself a social norm violation, providing an opportunity to learn more about how such technologies may be perceived and used, and to test and "trouble" our assumptions about the cultural categories that we presuppose in design, such as the role of user and bystander. In their description of field studies with mobile delivery robots in public spaces, Bødker shows how broadening the understanding of HRI beyond direct engagement can help us better understand who the stakeholders in robot use are, and how they might experience and affect robot use. Standing in my fenced-in garden on a Saturday morning with my husband and two children as we all stared up into the sky where a drone hovered, allowing some unknown other to observe our activities from above, certainly made us think differently about the boundaries of our private space and the new types of mobility and surveillance made possible by flying robots.

Velonaki's "experiential exhibits" (Box 3.8) provide another open-ended opportunity for people to interact with and experience new robotic interfaces, and for designers to learn from these interactions. These exhibits foreground the creation of novel forms of interaction through multisensory design features which include novel materials and relational movements. I recall spending time with her Fish-Bird exhibit in 2005, picking up and reading small scraps of paper with typed-out messages produced by two robotic wheelchairs as they attempted to communicate their inner thoughts. Moving around the space and trying to engage the slow-moving robots, which seemed both drawn to and repelled by both each other and the humans in their exhibit space, provided mental space to think about human connections with machines, machine interactions and people's connections with each other. It also provided a new perspective on the interfaces through which interaction can unfold – through small notes, small relational movements among several partners, a continuous dance of moving toward and away from, without a concrete resolution (aside from the exhibit's end, perhaps). Velonaki's work also suggests the importance of engaging with diverse aspects of experiential design – sound, movement, touch, and other multisensory features – to extract design principles.

Suchman (2007) has richly shown how important the study of "situated action" is in the use of computing technologies; the ability to incorporate it into a design is therefore crucial to creating systems that can be more usable and socially appropriate. Forlizzi et al.'s notion of "user enactments" (Box 3.1) displays how designers can systematically explore the potential use and consequences of certain robotic features in new contexts, allowing users to figure out how they might use and respond to technology in new use domains and incorporating novel features. Through evaluating imagined scenarios of robot use in vignettes, videos, and interactions "acted out" in various ways, users can provide socially situated responses to potential robot designs that can help designers explore new design features and application domains. Ocnarescu and Cossin (Box 3.4) further suggests the notion of "pretotyping" through low-tech prototypes, intention scenarios and low-tech living labs can provide users with a simulated experience of robot use that shows potential future interactions with robots without seeming to suggest that they are pre-determined outcomes, opening up the possibility for critique and amendments that are prompted by potential users. These sometimes dramatized, wizarded experiences can bring together the designers' imagination of potential futures with the situated user interpretation and experience of them, making it more clear what kind of mutual shaping might result and enabling more socially embedded decision-making on how to proceed with robot design.

Considering the social context of design, it is important to recognize that many contemporary robotics designers still come from WEIRD (Western, Educated, Industrialized, Rich, Democratic) backgrounds. The envisioned users of robotic technologies, though they can include people from similar backgrounds, come from a much broader swath of society, including those from less privileged backgrounds and in more vulnerable social positions (e.g., people with disabilities, older adults, children) as well as in countries that do not themselves create or produce robots. To enable more equitable and socially just solutions, the design process needs to contend with the power relations and differentials between the developers and potential users of robotic technologies, as well as other stakeholders that may be affected by the use of robots. They also need to incorporate these stakeholders to understand what matters most to them, bringing in issues like cost, socio-technical infrastructure, and other local concerns that affect the use and effects of technology.

One approach to this is seen in the method of "collaborative mapping of values with robots," described by Gonzalez and Jacobs (Box 3.2). This approach brings relevant stakeholders together to discuss values important to them in the use context, as well as the values of the developer and the company producing the robot (if relevant). Along with mapping values, they also consider how robot design features – behaviors, appearance, tasks – may fit those values to translate the abstract notion of values into more practical insights related to robot design. This approach addresses the different perspectives and needs of diverse stakeholders in robot design from the very beginning and provides a practical translation of those into robot design. Another collaborative design method, Lee's "Collaborative Map-Making" (Box 3.3), provides a way to explore the "matters of concern" that are priorities for participants, particularly those in underserved communities. It also allows participants to provide their perspective on the world – their framing of the situation – so that it can then create the foundation for design. Ensuing HRI design activities can then follow through by focusing on specific issues participants want to address, as understood by those participants – not just the researchers themselves.

"Co-design" can incorporate these types of activities and others in working directly with users on the design of specific robots. As described by Winkle (Box 3.5), co-design assumes equal authority of different participants in the design, including potential users and formal robot developers and designers. The mutual learning that undergirds this process can lead to changes in direction of project as well as guiding the specific aspects of the design itself, as users help define the terms of the project. These design methods provide ways to question and negotiate the values embedded in robot design from the earlier steps of the design process, before specific uses and expectations are already built in and pre-determined. Co-design can be supported by processes that allow users control over the design process, such as Senft's "end-user programming" (Box 3.7) – low-code or no-code programming that makes robot development broadly accessible – which provides a way for participants to directly instantiate their own ideas of what the robot needs to do and be. These methods seek to engage increasingly diverse populations in robot design, decentering the focus on robotics as a technical endeavor and recentering the human and societal dimensions of design as focal to the robot development process. End-user programming can also extend the design process into later use, allowing users to adapt the functions of the robot as they figure out how to make it fit their values, needs, and the patterns of everyday work and life.

Attention to the notion of mutual shaping extends the concept and process of design to include everything from the moment of initial conception of a design problem, through its ideation, implementation, subsequent evaluation and iterations, and finally its adoption and adaptation by users until the technology is no longer in use. The methods in this section provide the tools needed for designers, developers, potential users, and others who stand to be affected by robots, to participate in the visioning and creation of our future experiences with these emerging technologies.

BOX 3.1 USER ENACTMENTS

Jodi Forlizzi, Carl DiSalvo, Bilge Mutlu, Min Kyung Lee, Michal Luria, Samantha Reig, and John Zimmerman

User enactments explore how to design technology in novel contexts. This method allows design researchers to prototype narratives and social contexts through simulated scenarios or roughly prototyped representations of products that do not yet exist. Research participants are asked to read storyboards or to enact loosely scripted but familiar social situations where new, unfamiliar technology systems are situated. Participants then reflect on what a preferred future is for them and why. This method is relevant for HRI research because it allows a research team to evaluate an ample design space, using storyboards or partial prototypes with some portions working and others as simple demonstrations, to **explore how technology might be situated** in a physical and social context.

For example, Lee et al. (2010) employed user enactments to test four recovery strategies – apologies, compensation, options for the user, and doing nothing – to mitigate robotic breakdowns. The method can also be used to design for populations that are difficult to study in the lab and when the physical and social context plays a crucial role in using the technology. Stegner et al. (2023) engaged target user groups in co-designing experimental materials. The research team can develop rapid prototypes, combining autonomous and Wizard-of-Oz capabilities to support these scenarios. Structured user enactments follow a more defined script. When researchers need to systematically explore a design space, the structure of the scenarios may appear similar to conditions in a controlled design (2×2, 2×3, etc.). Once initial design research directions emerge, structured user enactments are better used in HRI. Reig et al. (2020) used structured user enactments to explore how robots should interact with individual users appropriately in a social setting.

AN APPLICATION OF USER ENACTMENTS

In 2010, as research robots were becoming robust enough to consider using them for studies in the real world, our lab regularly witnessed robots in the real world making mistakes. We wondered what the best ways to mitigate those mistakes might be. A robot's behavior might be designed to answer this question in thousands of ways.

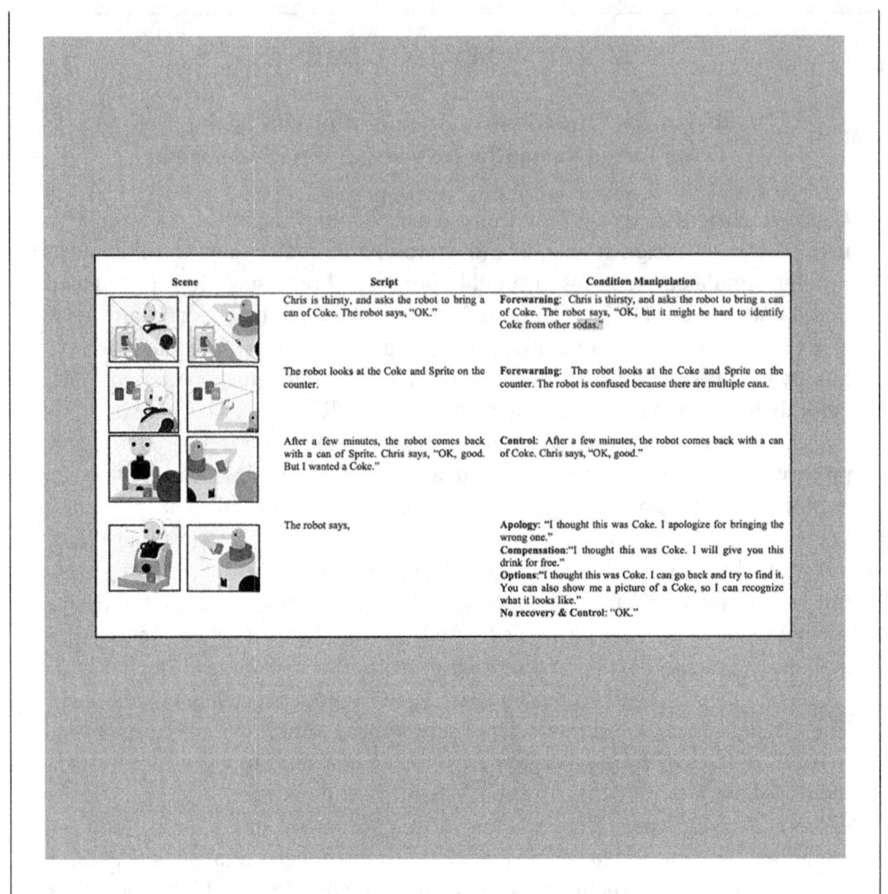

Scene	Script	Condition Manipulation
	Chris is thirsty, and asks the robot to bring a can of Coke. The robot says, "OK."	**Forewarning**: Chris is thirsty, and asks the robot to bring a can of Coke. The robot says, "OK, but it might be hard to identify Coke from other sodas."
	The robot looks at the Coke and Sprite on the counter.	**Forewarning**: The robot looks at the Coke and Sprite on the counter. The robot is confused because there are multiple cans.
	After a few minutes, the robot comes back with a can of Sprite. Chris says, "OK, good. But I wanted a Coke."	**Control**: After a few minutes, the robot comes back with a can of Coke. Chris says, "OK, good."
	The robot says,	**Apology**: "I thought this was Coke. I apologize for bringing the wrong one." **Compensation**:"I thought this was Coke. I will give you this drink for free." **Options**:"I thought this was Coke. I can go back and try to find it. You can also show me a picture of a Coke, so I can recognize what it looks like." **No recovery & Control**: "OK."

FIGURE 3.1 A service robot storyboard as the basis for enactment.

To explore this large and complex design space, we looked at service literature on expectancy-setting strategies (Smith et al., 1999) and recovery strategies (Kelley et al., 1993), including apologies, compensation, and providing options for the customer. We used two different robots in our study because we wanted to understand if the robot's design had any impact on participants' interpretations of its behavior. We then designed a speed dating study in the form of storyboards which explored the following scenario:

> Chris is thirsty and asks the robot to bring a can of Coke. The robot says, 'OK.' The robot looks at the Coke and Sprite on the counter. After a few minutes, the robot returns with a can of Sprite. Chris says, 'OK, good. But I wanted a Coke.'

Following the service literature, the robot then did four different things: 1) nothing (no recovery), 2) provided an apology, 3) offered the Sprite for free; 4) offered to go back and get the Coke instead. In addition, in half of the scenarios, the robot warned that the task was hard and it might make a mistake (forewarning).

Three hundred seventeen participants reviewed these scenarios online. They first looked at a short video of each robot, then examined the storyboard scenarios, evaluating the robot, whether it gave excellent or poor service, and how likely it would be for the character in the scenario to use the service again. Overall, expectation setting was the most effective in mitigating the negative evaluations of the robot. All of the recovery strategies increased positive ratings of the robot's politeness. The apology and option strategies effectively improved the participants' perception that the character in the scenario would use the service again.

This was a novel approach to exploring an inordinately ample design space of options and providing information about designing robots that can fail gracefully.

BOX 3.2 ROBOT VALUE MAPPING

Irene Gonzalez and Jan Jacobs

Robot Value Mapping leverages collaboration to define and incorporate context-specific values into the design of HRI. The method builds upon Value-Sensitive Design (VSD), pioneered by Batya Friedman and her team (Friedman et al., 2002). VSD provides theories and techniques for the design of technology that account for human values. However, this practice ought to be modified to adapt to different values, technologies, and contexts of use. Notable examples of such adaptations include the frameworks developed by Steven Umbrello and Ibo Poel (Umbrello & Poel, 2021) for AI, as well as the contributions of Aimee Van Wynsberghe (Van Wynsberghe, 2013) in the realm of care robots. Robot Value Mapping, thus, implements VSD in the design of robot behaviors through collaborative mapping sessions. These mapping sessions can serve various purposes. They can aim to identify the desirable values to underlie a robot's behavior, such as "reliability," "integrity," or "humor." This goal includes **capturing the explicit meaning of these values within the specific context** in which the robot operates.

Furthermore, the sessions can focus on establishing processes to guide the development of robot behavior based on these identified values. Robot Value Mapping can also be used to assess current robot behaviors and the resulting HRI, identifying areas for improvement and enabling iterative enhancements based on them. To effectively recruit relevant stakeholders for these sessions, it is vital to have a clear purpose. The mapping activities can then be tailored accordingly, taking into account the goal as well as the profiles of the participants. These activities foster meaningful discussions, uncover potential frictions, and facilitate the synthesis of ideas. Empowering stakeholders to utilize examples when discussing ambiguous concepts, such as values, can help prevent miscommunication and enhance comprehension. The adoption of Robot Value Mapping not only helps companies prevent unintended harm and ensure ethical alignment but also enables them to differentiate themselves by intentionally showcasing their brand identity and value proposition through robot behaviors. Users, in turn, benefit from value-grounded robot behaviors by gaining transparency, predictability, and personalized experiences that cater to their specific needs.

ROBOT CODE OF CONDUCT

We embarked on a project aimed at integrating HRI knowledge into the development practices of a multinational automated dairy farming company (Gonzalez, 2022). This company, like many others, has established and respected codes of conduct orienting their employees' behaviors. Our goal was to define the values that would inform the behavior of their extensive robotic

FIGURE 3.2 The table used to host a mapping session.

portfolio. Determined to **go beyond aesthetic design guidelines**, we sought to provide a normative representation of robot values and behaviors tailored specifically to the targeted company. This would facilitate mindful, efficient, and cohesive decision-making throughout the robot design process. We employed collaborative value mapping to assist a representative sample of experts in identifying these robotic values. The collaborative value mapping session took the form of an onsite workshop at the company headquarters. The participants came from various departments, including product development, customer advisers, and representatives from the organization as a whole. To ensure a **holistic understanding of the context** among all participants, we began the session with a sensitizing activity and an interactive presentation. It is highly complex to envision and articulate robotic values, so we designed the activities following the path of expression. Participants started mapping their own behaviors and values in their respective jobs using large open templates before exploring how robots could embody and represent different values. For the latter, they embarked on a role-play exercise in small groups where a "robot," "farmer," and "observer" would enact and analyze various interactive scenarios. Through the mapping exercises and subsequent group discussions, we gathered rich insights into stakeholders' diverse perceptions and preferences. Concrete examples were used to convey individual interpretations, enhancing the clarity of discussions. We compiled those insights into a list of robotic values. However, we wanted to go beyond abstraction and ensure that these values would effectively influence design decisions. To achieve this, we conducted a second collaborative value mapping session with a different focus. This time, we aimed to investigate how these values resonate with the development teams and could guide the development practice.

Therefore, we involved participants representing different robotic product teams, such as hardware, software, and mobile applications. We adapted the session to the new goal and audience. This time we followed a more structured approach. Instead of open templates and roleplaying, we mapped the categories of the company robots and the proposed robot values in a 9 × 8 cell table and used it to host all session activities. The participants first familiarized themselves with the values and explored the extent to which they materialized in each robot category of their portfolio. Then, they ideated future robot behaviors rooted in these values, concluding with a discussion on how the acquired knowledge could inform their practices. Analyzing the outcomes of the second workshop, we refined the list of robotic values. Moreover, **we learned how these values could be effectively conveyed to different audiences** and took initial steps toward the design and implementation of robot behavioral guidelines. Collaborative value mapping, however, benefits from careful considerations, extensive multidisciplinarity, and periodic iterations.

BOX 3.3 COLLABORATIVE MAP-MAKING

Hee Rin Lee

Collaborative Map Making is a method that helps design researchers address previously invisible issues from the perspectives of participants, especially those who are from underserved communities, whose knowledge has been devalued, and whose voices have been weak in HRI. This method involves the combination of mapping and collaborativity, thus the name. Mapping involves researchers creating maps of the situations they study, based on what matters, rather than on existing rules, frames, or practices.

The theory underpinning this aspect involves Latour's "matters of care" (Latour, 2007), in that Collaborative Map Making enables researchers to examine their issues by constructing their own networks. These networks are composed of relationships of various actors, both human and non-human; these actors are typically considered to be irrelevant when following existing frameworks but are closely related to each other when prioritizing what matters. For example, according to the conventional framework of assistive robots, "aging" is most closely related to "losing capabilities"; however, when older adults map out their aging experience, what they consider more relevant is the "wisdom" they accumulated through empirical experiences.

The practical aspect of mapping is inspired by Clarke's situational map making (Clarke et al., 2017), which focuses on analyzing relations among various actors (Lassiter, 2005). Collaborativity is derived from Lassiter's collaborative ethnography (Lee et al., 2017), in that both methods **re-position participants from being informants, to being co-knowledge makers**. This addresses the power imbalance between researchers and participants. Considering the devaluation of lay people's contextual knowledge when compared to computational/technological knowledge in HRI, this collaborative aspect emphasizes that participants are not merely co-designers. HRI researchers can use this method as a first step to collaboratively define what issues participants want to address. More details on Collaborative Map Making can be obtained from (Lee et al., 2017).

REFRAMING ASSISTIVE ROBOTS WITH OLDER ADULTS

Although older adults are one of the most studied populations in HRI, they have limited chances to share their perspectives. Additionally, their aging experience is framed as a process of losing capabilities, which robots are expected to compensate for. To challenge these stereotyped views of older adults, I conducted collaborative map making studies with older adult participants in the US (Lee et al., 2017; Lee and Riek, 2018; Lee and Riek, 2023). In those studies, older

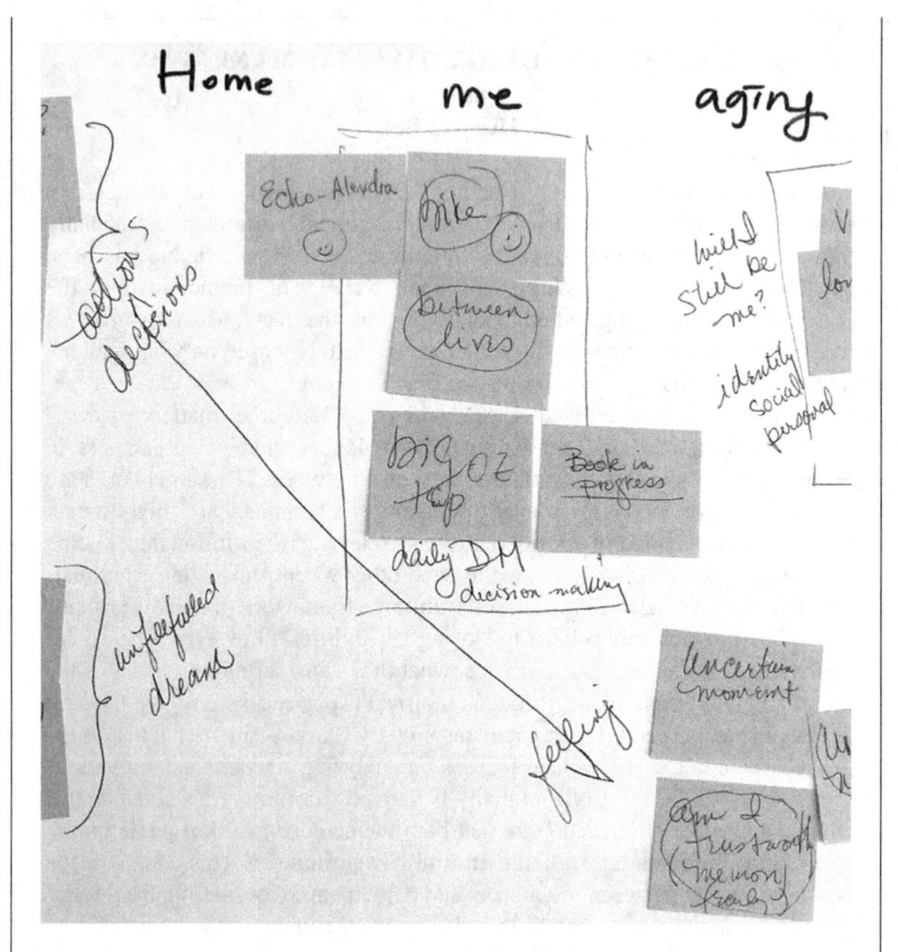

FIGURE 3.3 *A map created by an older adult.* An older adult participant came up with five groups: (a) daily decision-making, (b) will I still be me (identity), (c) feeling, (d) unfulfilled dreams, and (e) continuous decisions. The circled post-its were the most important words for her, which belong to either the "daily decision-making" group or the "feeling" group. This participant enjoyed thinking about the relationships, between groups and actively visualized them on her map.

adults mapped out their aging experiences from their own perspectives. By using **Collaborative Map Making**, I – an HRI researcher who has learned previous frameworks on assistive robots, and who has not experienced aging in the way that older adults have—tried to avoid framing older adults from my own perspective, which was based on existing HRI research practices and frameworks. I went through the following steps with each participant.

[Step 1: three keywords] I chose three keywords that I am interested in, which were "home," "aging," "you" in the study with older adults.

[Step 2: 15 words] After presenting the keywords, I asked participants to write a total of 15 words related to the three keywords, with each word on an individual Post-it.

[Step 3: most important words] Next, I asked them to choose the most important words (1 to 3 words) of their 15, and to mark these with a pen.

[Step 4: categorization] Then, I asked them to categorize those 15 words into 3 (or more) groups, and to give each group a title.

[Step 5: relationships between words] Participants were also encouraged to visualize the relationships between words or groups (e.g., a line connecting two words to show their influence on each other).

[Step 6: map explanation] When the mapping was completed, participants were asked to explain the meaning of each group and word, the relationships that they marked, and the reasons behind ranking words as important. I also asked questions to participants to make sure I was interpreting their intentions correctly.

The precise methodology involved in each step is flexible, and can vary if participants experience difficulty; what is important is the underlying concept. The conversations were recorded, transcribed, and analyzed. The maps were also analyzed together with conversations. Both sources of qualitative data were analyzed through a grounded theory approach. In older adults' maps of aging, the main issue was never losing capabilities. Rather, they wanted to develop resilience to better adjust themselves to the changes in their lives and reinforce their existing capabilities. The maps also **revealed that the stereotyped views toward aging** caused older adults social, economic, and psychological challenges; these challenges could be amplified by robots developed upon stereotyped representations of aging. Because building or testing a robot takes much effort and time, HRI researchers can often miss the opportunity to genuinely understand the stakeholders' perspectives. Collaborative Map Making will enable HRI researchers to focus on researcher-defined issues to participant-defined ones. This method is especially beneficial when studying stakeholders from marginalized communities whose knowledge has been historically undervalued, such as union workers (Lee et al., 2023) or family caregivers (Shin et al., 2021).

BOX 3.4 PRETOTYPING

Ioana Ocnarescu and Isabelle Cossin

It is sensible and challenging to design robotic solutions for people in real-world environments. Several design tools could help explore potential cohabitations with robots from field investigations, taking into account the ecology of living for reeducation patients, and before building functional prototypes. These investigations can be carried out in a variety of ways, including observation through sketching, concept development, formal prototypes, and design interventions that put these design prototypes into use. These are designerly ways of **pretotyping**.

The concept of pretotyping was proposed by Savoia and colleagues (2011) as part of their work at Google. The goal of pretotyping is "*to explore and test, quickly and cheaply, extreme ideas that would normally be dismissed as too risky or expensive to try, … to make sure that you are building The Right It before you build It right.*" Design professionals use experience prototyping and embedded design methodologies to test ideas and concepts. Design pretotypes are tangible hypotheses and intermediate representations with the ability to generate knowledge and bring a tangible focus within the design process. Within this exploratory and embedded approach for HRI, we suggest several tools: **low-tech prototypes, intention scenarios**, and **low-tech living labs**.

The three design tools will be further discussed in the context of an HRI project focusing on assistive robots. They made a contribution to the development and investigation of use-cases and scenarios involving assistive robots for elderly people. They served as intermediary representations in deciding which technical solutions are desirable to use when creating an assistive robot. Moreover, better assistance is not just about developing functional tasks that work effectively; it must consider the overall experience of assistive technologies in people's lives. The knowledge produced by these tools goes beyond this specific technical requirement of "what use-case to implement." They also support the implementation of comparative studies of *in situ* and *in vivo* HRI investigations. These design tools coupled with both qualitative and qualitative research gave direction and methodological insights to better capture the complexity of social robotics in real settings.

ROMEO2

A 140 cm tall humanoid robot named Romeo was the goal of the Romeo2 project (Ocnarescu and Cossin, 2017; Ocnarescu and Cossin, 2020), an HRI research consortium led by the French research institutions. Romeo was intended to help dependent people living in French nursing homes and rehabilitation centers. An HRI team comprised of doctors, researchers, and designers was established to imagine typical scenarios and practical use-cases involving Romeo. The formal prototype of Romeo already existed; the functional one

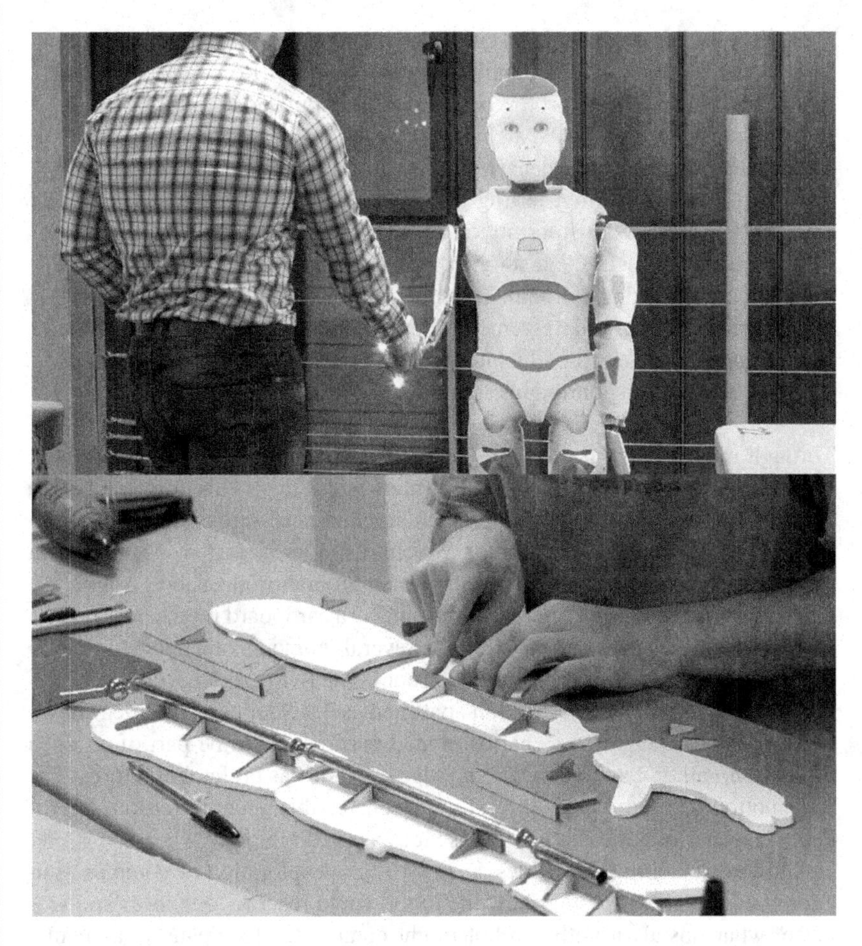

FIGURE 3.4 The making of a paper pretotype of the robot Romeo and its use.

was under construction at Softbank Robotics (now Aldebaran). Following a phase of observation in a number of rehabilitation facilities, designers proposed Anubis, a different robotic assistance that was not humanoid. To explore new concepts of social robots for assistance we used low-tech prototyping techniques in two different formats: intention scenarios and low-tech living labs. With the help of the various pretotypes produced by this novel approach, these design interventions contributed to the consolidation of several studies.

Low-tech Prototypes: Low-resolution skeleton-type prototype can be built using basic material from a Fab lab (3D printing machines, laser cutters etc.). The figure below shows different ways to building a 3D form. We chose a mixed technique consisting in 3D printing the small parts and coupling them with metallic tubes to maintain the structure.

We used a cardboard articulated with metallic tubes and magnets on the back to build a tangible representation of Romeo robot. Both Anubis and Romeo prototype could be manipulated in terms of movement. We created 1:1 scale models for the two prototypes.

Intention Scenarios: Scenario prototyping is a continuous process of exploration, making, testing and discussions. We used Video-based HRI (VHRI) with stop-motion animation for new exploratory investigations. VHRI tells a story in a context and it unlocks tacit knowledge observed in the field. To build a stop-motion scenario the object is moved in small increments between individually photographed frames. It creates the illusion of movement when the series of frames are played as a continuous sequence.

Through movement the prototype will acquire a character (even archetypal). We called these video prototypes intention scenarios. This tool is also an appropriate to study people's appreciations, and perceptions on robot's personality, appearances, and behavior. We build similar intention scenarios with three robot prototypes: Nao – a small size anthropomorphic robot, Anubis – a non-humanoid actuated skeleton, and Romeo - a cardboard representation of a humanoid robot. The scenarios describe several moments of a person interacting with an assistive robot in her living room. The living room decor was carefully chosen in order to look like an apartment of an elder person. When shown to an audience, intention scenarios invited participants to be part of an open discussion, rather than making them judge scenarios as preferable futures. The stop-motion technique was carefully chosen for this specific reason. It proposed a scenario, but it did not impose it as a certain future.

A **low-tech living lab** enables an *in vivo* HRI exploration of complex situations and emotions, such as how the robot would react to loneliness and violence, what nostalgia with a robot might entail, etc. To create an authentic living environment, we set up a one-bedroom elderly apartment filled with furniture and other items that we collected from the homes of the students' grandparents. We created stories—sometimes dramatic, sometimes provocative—that explored various hypotheses. Like in a theater play, we interacted with two Pepper robots that were animated using Wizard of Oz puppeteering techniques. We spent one full week, from morning till night with medical staff, ergotherapists, designers, engineers, and the two robots. The scenarios were lived, played, and filmed by the persons taking part in this experience. This *in vivo* experiment allowed us to explore the interactions with the robot at different scales: to think the gestures and the micro-interactions between the participants and the robot, but also to explore more complex scenarios. At the end of the week, we obtained 20 videos describing different life and a document with photos and recommendations on the subjective experience of the participants. The technical advancements on an autonomous robot for an *in situ* experimentation in rehabilitation centers (24 participants living with an autonomous robot for one week) were chosen using this information.

BOX 3.5 CO-DESIGN

Katie Winkle

Co-design is about working directly with users on the design and development of your robot - whether that's how the robot looks, how it acts, or even how it is intended to be used. Compared to user-centered, and participatory design, a distinguishing feature of co-design is that participants should have equal authority to the designers, driving the overall what, why, and how rather than working to deliver on designers' pre-defined agenda. Key to achieving this equal authority is the process of **mutual learning**, a two-way exchange of knowledge and ideas between participants and designers. Designers educate participants about the technologies they're working with, which helps ground participants' ideas in how to use them. Participants educate designers about the requirements and realities of their application domain. Together then the participants and designers come together as one design team.

Co-design implies an overall approach rather than any one specific methodology; common activities used to support co-design include focus groups, prototyping workshops, and storyboarding, generally combined with some sort of educational element designed to help participants get to grips with the technologies under discussion. This could take the form of designer-led presentations and demonstrations or more hands-on activities such as programming exercises. To achieve the **meaningful engagement of/shared authority with users**, co-design typically requires multiple designer–participant meetings. Similarly, the design project can often go in unexpected or unforeseen directions – something worth thinking about when planning your own design project. If your design or research agenda is relatively fixed, with limited potential for participants to really shift the project's direction, it might be worth considering whether participatory design might be a better fit for your work. The methodologies and mutual learning processes described here support this equally well, but participant involvement is generally limited to ideation and refinement rather than project direction.

EXPLORING ROBOTS FOR GROUPS WITH TEENAGERS

A number of social robotics research projects have explored the potential for robots to mediate group interactions between children and young people – using robots, e.g., to manage turn-taking, and ensure everyone gets equal time to

FIGURE 3.5 Participants generate multimodal robot actions using a design template. The template includes a multi-colored hexagon where each section of the hexagon identifies a communication modality to remind designers to reflect on which modalities they wish to use, and how, for a particular action.

speak, and guide play activities designed to improve inclusion between children of mixed ability or sociocultural background. Where these works generally use literature to specify the "problem" and inform the possible design of a "solution," we instead employed a **co-design approach** to explore robots for teenage groups with teenagers themselves. We conducted a two-week "summer school research study" with an activity schedule designed to support the mutual learning and co-design aims of our work (the research study part), while also providing an enjoyable learning experience for our participants (the summer school part).

Our design brief to our participants was minimal – design a group robot assistant that can make group working *better*. Our first task was therefore to work together with participants to identify what *better* meant for them, and hence set some design specifications for the robot accordingly. We did this via focus group discussions in conjunction with group communication games and activities. At this point, after introducing some previous literature on social robots for groups and undertaking some (roleplay) "robot programming" activities designed to demonstrate the need to design multimodal robot behaviors, we moved on to iterative robot action design, where participants ultimately identified a set of actions for our group assistant robot.

We the design team, implemented participants' actions into the robot and produced a tablet interface through which participants could execute these actions on the robot in real-time. Following a final refinement of action design based on a first session using this setup, participants engaged in a series of robot group working sessions wherein group members rotated between controlling the robot, working with the robot, and witnessing and evaluating the group–robot interaction.

This setup was again designed to ensure **participants took the lead** in dictating how the robot should best support their group working. As we explained to participants, our ultimate aim was for these sessions to represent "robot training sessions," during which the robot would learn from them, via interactive machine learning, how best to behave. In this way, we identify interactive machine learning and participatory automation as another tool to support robot co-design. The robot training sessions were preceded by, and interspersed with educational activities introducing, e.g., supervised machine learning and the potential for bias, designed simultaneously to ensure participants understood our notion of "robot training sessions" but also to support our aim of also providing a valuable learning experience beyond the short-term aim of our research study.

BOX 3.6 BREACHING EXPERIMENTS

Mads Bødker Rosenthal-von Der Pütten

Robots are assumed to soon become part of public life, and HRI researchers need to study how people understand and cope with the presence of robots in everyday life. The term **Incidentally Co-present Persons**, *InCoPs* (Rosenthal-Von der Pütten et al., 2020) describes the "passive" bystanders to robot activities, for example, bystanders to delivery robots or garbage collecting robots in public spaces. Studying InCoPs in-the-wild (Crabtree et al., 2020) through customized "breaching experiments" and membership classification analysis can provide HRI designers with cues to shape the proper conduct of robots in natural and uncontrolled settings, ensuring that robots are perceived in a meaningful way.

We take a cue from Harold Garfinkel, who described "**breaching experiments**" as simply "making trouble" within a familiar scene (Garfinkel, 1967), to expose contextual assumptions about meaningful or acceptable behavior in a setting. In this case, the "trouble" is simply the introduction of a robot in a familiar setting (say; a pedestrian area, a sidewalk, a school atrium, a park) which "provokes" bystanders (who are naïve to the study) by being in some way intrusive or unusual. The robot can be remotely controlled by a "wizard" researcher to create the illusion of an autonomous robot.

The researchers can freely imagine what kinds of robot behavior will be considered provocative or striking within the context, keeping in mind participants' safety and local customs or ethics. A researcher is placed to observe people's reactions, and, when possible, conduct short interviews or debriefs. Of special interest for researchers is "membership classification," where the "members" of the social context are asked how they would categorize the robot (e.g., as a threat, a toy, or a tool). A light-weight method, the breaching experiments with InCoPs is easy to run. They **expose local norms and perceptions**, providing cues for the design of contextual robot behaviors.

INCIDENTAL ENCOUNTERS WITH ROBOTS

Curious about the relation between public space, robots, and "non-users" (that we labeled InCoPs), we (two students and the author) applied the method in a study conducted at an outdoor pedestrian zone in Copenhagen in 2020 (Moesgaard et al., 2022). The walkway connects two main buildings of

FIGURE 3.6 The robot making trouble in the pedestrian zone at Copenhagen Business School (CBS).

Copenhagen Business School (CBS), and we used "Wizard of Oz" simulations using a remote-controlled outdoor mobile robot to create different unusual situations (or "provocations," breaches of ordinary protocol in the context) that we found would occasion people (the bystanders/InCoPs) to react and reflect. These **provocations** were unsystematically constructed, based on vague assumptions (as well as technical limitations to the robot), and included having the robot stuck and trying to get out, approaching people from the front, behind, or from the side, driving with abrupt and unpredictable behavior, standing still in the middle of the pathway, using sounds to get people's attention, the robot blocked by an e-scooter, or "dressed up" as a mobile heart defibrillator. During the study, 338 encounters were recorded. Participants were recruited for short interviews *in situ* as they incidentally passed by the robot. No prior information about the experiment was given to the participants, and none were rewarded.

The informants were grouped by their age (adult, child), their self-reported status (student, senior citizen, caregiver with children, cyclist), and asked how they felt about the robot that they had noticed. For the analysis of the interviews, we used a loose interpretation of the concept of "membership categorization" from **ethnomethodology**, trying to understand how people categorize the robot that had made a bit of "trouble." Indeed, it seemed that more "provocative" behavior by the robot (driving erratically or being stuck and trying to get out) improved the richness of the follow-up interviews and made it easier for interviewees to articulate how they perceived the robot and how they would classify it. Based on their different demographics and working knowledge of the context, people classified the robot as entertainment, a threat, a work-related robot, a pet, an experiment, etc.

Our study was relatively unstructured and can perhaps best be described as an early exploration of "breaching experiments," "membership classification" and other ethnomethodological concepts that we found intriguing for the study of robots in-the-wild. We emphasize the way in which relatively unstructured in-the-wild studies can be used as "aids to a sluggish imagination" (Garfinkel 1967: 38), particularly in the early stages of the development of robot behaviors within a particular context.

Understanding the in-the-wild conviviality between place, human and (non-human) robot actors constitutes a continuing challenge for HRI research and practice. The main findings of the study point toward the "experiential" and "aesthetic" aspects of robots. Rather than measuring efficiency or other productive aspects performed through HRI, our study speaks to the assumption that understanding how "bystanders" perceive, and experience robots is important, as future public spaces may plausibly be busy with both human activity and autonomous robots.

BOX 3.7 END-USER PROGRAMMING

Emmanuel Senft

End-user programming (EUP), sometimes referred to as low-code or no-code programming, is a development approach aiming at streamlining and simplifying programming for users without a formal programming background. In the context of robots, its main goal is to allow end-users, who are often **non-experts in programming**, to program or reprogram robots, see (Ajaykumar et al., 2021) for a recent review in the context of robotics. EUP methods can use a variety of program representations, including: (1) linear programs with sequential actions (e.g., (Steinmetz et al., 2018)), (2) flow-based programming which supports branches and loops (e.g., (Pot et al., 2009)), (3) behavior trees (e.g., (Paxton et al., 2017)), (4) trigger-action programming which represents robot actions in an asynchronous way (e.g., (Senft et al., 2021; Leonardi et al., 2019)), and (6) behavior cloning where a robot can replay a behavior (e.g., (Fang et al., 2019)).

To be able to create, edit, debug, and verify programs, researchers have developed a **wide range of methods and interfaces**. Visual programming shows the program in a graphical way, allowing users to move blocks of code (Coronado et al., 2020). Wizard menus can allow users to progressively select variables and parameters of their program. Users can also utilize annotations and drawing, representing their program on a canvas, or use tangible interfaces to annotate the physical world, or mixed reality overlaying virtual interfaces on the physical world. Programs can also be represented and edited through timeline and keyframing to synchronize multiple actions over a period of time, for example, by manipulating the robot directly, through kinesthetic teaching or puppeteering. Finally, research has also explored how verbal interfaces could be used to program robots.

AN APPLICATION OF EUP IN AVIATION MANUFACTURING

EUP is becoming common in manufacturing environments as a way to allow factory workers to **repurpose or customize robot behaviors on site**, without having to rely on an engineer to update the robot code (Rossano et al., 2013). For example, in the context of aviation manufacturing, sanding is ubiquitous, almost every piece used in an aircraft has to be sanded at least once. However, sanding can be straining on workers and potentially lead to injuries during excessive sanding periods. As such, automation through robotics is an attractive solution to protect workers, but it may be hard to justify the investment required to develop fully autonomous sanding solutions.

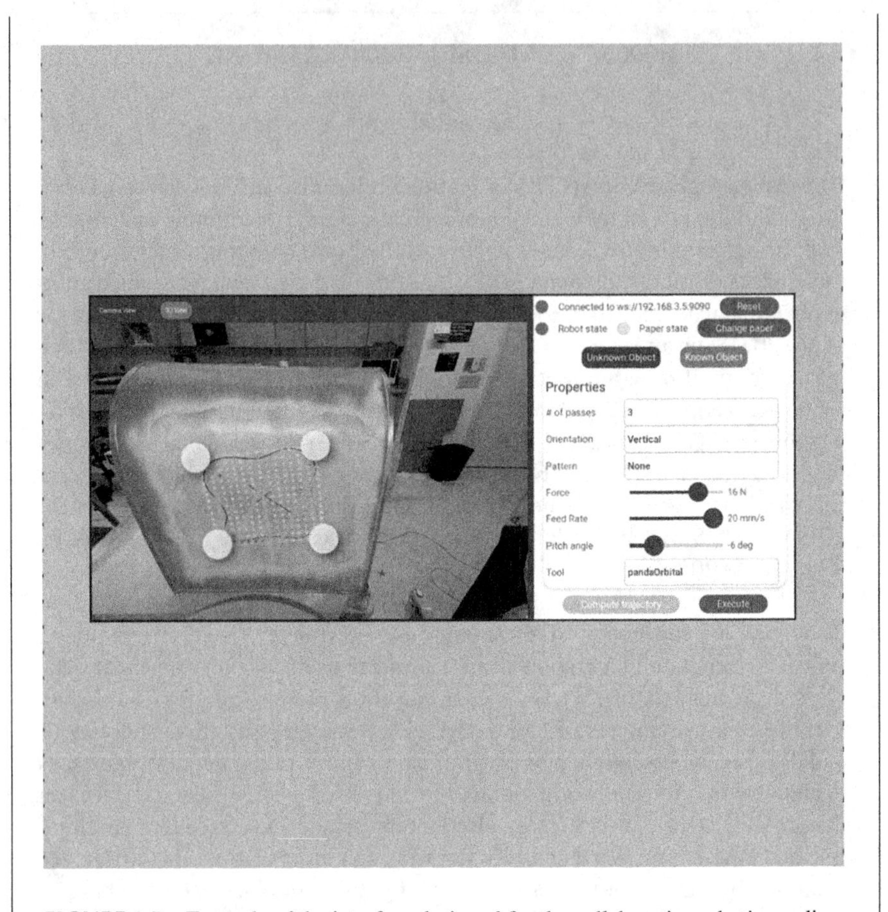

FIGURE 3.7 Example of the interface designed for the collaborative robotic sanding.

As part of the NASA ULI project *Effective Human–Robot Teaming To Advance Aviation Manufacturing* a team at the University of Wisconsin–Madison collaborated with a major aviation manufacturer to develop EUP approaches allowing shopfloor floor workers to design sanding robot behaviors on the fly. One of the main assumptions for this work is that no accurate model of the piece (e.g., sanding of the barrel to change the coat of paint) is available. Consequently, the system needs to perceive the environment, design a plan using the operator's instructions, and execute it. A final challenge is that due to the reliance on local perception, the robot plan might be suboptimal, as such, we need ways to maintain the user in the loop to ensure success in the task.

In practice, the operator starts to position the robot and select on the fly the area to be worked on, visually verify the reachability of the robot, and finally specify the sanding characteristics. To maintain the user in the loop during the execution and benefit from their expertise and vision capability to increase the robustness of the system, **the operator could provide corrections** to the robot's trajectory and forces.

The final system was composed of a Franka Emika robot outfitted with a Kinect Azure, a force torque sensor, and a custom end effector using workers' manual sanding tool to support better transfer of their expertise.

The EUP interface was inspired by (Senft et al., 2021) and was displayed on a web browser application running on a phone outfitted with joysticks to support real-time corrections. The figure shows an example of the interface used, the left pane of the interface shows the view from the robot-mounted camera that workers can annotate to select the sanding areas. An overlay also shows the reachability of the robot as well as the computed path according to the sanding parameters (e.g., applied force, tool pitch, feed rate…) that are specified on the right pane of the interface.

With the selected area and sanding parameters, the system generates a trajectory using the depth channel of the RGB-D image that can be visualized in the camera view. And finally, the robot executes the trajectory using a hybrid control to maintain an appropriate force orthogonal to the surface. During execution, the user can use the phone's joystick to provide corrections along the trajectory (Hagenow et al., 2021), to address situations not handled by the EUP interface, such as a localized defect.

BOX 3.8 INSTALLATIONS AND PERFORMANCES

Mari Velonaki

Designing **installations and performances** enables the investigation of the "physicality" and communication of robots outside the laboratory. Especially when brought to public social environments where people can interact with robots more closely and for extended periods, installations, and performances generate a rich setting for designerly research entangled with the liminal, the quotidian, and the social (Murray-Rust et al., 2024). In these practices, **technical development goes hand in hand with aesthetics and cultural referencing** (Velonaki et al., 2008), opening up to a broader exploration of robot expressivity and communication. Much of the work on communication in HRI is traditionally focused on explicit modes of robot communication such as speech, bodily gestures, and, sometimes, text and other info channels. Performances and installations, however, encompass a broader set of features to convey expressivity, such as proximity, kinetic language, robot sonification, affective touch, cultural referencing, and more. Implicit modes of communication are here as important as the explicit ones and are explored as functional channels to convey presence and agency, and to give space to the enactment of possible human–robot relations and their emerging meanings (Gemeinboeck, 2021). Some of these implicit modes, such as movement and affective touch have been studied individually (see the work by Silvera-Tawil et al. (2014)), yet the combinations of implicit communication modes, and particularly on communication through sonification or scent emission remain a relatively uncharted territory.

Finally, exploring and designing HRI through installations and performances opens the field not only to a broader set of features and expressive channels, but also, and foremost to a **diverse set of practices coming from non-technical disciplines** that usually remain marginal in robotics, such as the theater, dance, craft, and other artistic practices.

DIAMANDINI REVISITED

Diamandini (Silvera-Tawil et al., 2015) is an interactive humanoid robot that responds to the movement and touch of an interactant. The robot's appearance is that of a **sculptural figure with a porcelain-like finish**, introducing a new aesthetics in robotics that is far removed from that of a stereotypical humanoid robot. To achieve the ethereal gliding motion required for Diamandini a novel omnidirectional motion base was invented and developed. The motion base has a patented mechanism to give the pure rolling motion of the wheels when the motion base changes velocity or angular orientation. Here I will present the three main modalities that were considered in creating Diamandini: Movement, Touch, and Sound.

FIGURE 3.8 "Diamandini" (2009–2013) by Mari Velonaki. Robotic installation. (Photograph by Paul Gosney.)

Movement

My work toward developing a kinetic language for robots has shown that how a robot moves, occupying and sharing space with a person, reveals a robot's intent and agency. Further, information in the form of movement is readily interpreted by people. Further investigation of robot movement in HRI is necessary to develop a qualitative kinetic language for enhancing robot physical presence, intent, and communication via expressive movement. For example, a robot should use movement cues when negotiating the sharing of physical space with people through harmonious blending with human movement patterns. Another knowledge gap is the lack of consideration of the pose in which a robot should stop, and how this pose could help to better communicate an interactive narrative. The kinetic embellishment of motion between robot poses to enhance a robot's agency and liveness has also been investigated.

Research in robot movement as a kinetic language started by my then PhD student Adrian Ball (Ball et al., 2015) by examining the degree of comfort that people have with the approach of a robot from different directions in a room. My previous work has been extended to include the effects of 1. different speeds, 2. changes in speed during robot approach, and 3. the impact of robot scale and representation.

Touch

There is currently very strong research and industrial interest in novel electronics for implementing touch sensing that is distributed across a surface.

In work with my PhD student and then Postdoctoral Fellow David Silvera-Tawil, electrical impedance tomography was used to create a soft, flexible, touch-sensitive artificial skin and to demonstrate that this skin could be used by a robot to identify emotions and social messages via classification of touch (Silvera-Tawil et al., 2014) at 86% of the accuracy achieved by human classifiers. Touch is closely linked with the materiality of a robot's exterior surfaces. In the context of "Diamandini," my interest is in the interpretation of touch on "its" surface.

Sound

The medium of sound is central to how humans perceive and communicate with their environment. While the communicative potential of sound is well-established in other disciplines, its application in HRI has to date been limited to speech and nonverbal utterances. Robot sound, however, occurs in a multitude of other contexts in HRI, where it has demonstrated effects that are much less understood. It influences people's expectations toward robots through motor noise and re-contextualizes the perception of robot movement (Robinson et al., 2021). It is performed by robots to engage and communicate through music and fosters robot-human relationships through shared listening experiences.

The sonification of Diamandini has been extended by collaborating with Frederic Robinson, my former PhD student, whose innovative work in emitting distributed sound from a robot as an indication of intent and affect (Robinson et al., 2021) allowed for effective communication from a robot to a person. Currently, we are working on enhancing Diamandini's communication by utilizing time-varying acoustic sound fields to generate the impression of sound movement in the proximity of a person. Sound will be emitted from spatially distributed sources which are integrated into a robot's body and operate in phase synchrony to evoke the impression of the robot's intent and effect.

REFERENCES

Ajaykumar, Gopika, Maureen Steele, and Chien-Ming Huang. "A survey on end-user robot programming." *ACM Computing Surveys (CSUR)* 54.8 (2021): 1–36.

Ball, A., Rye, D., Silvera-Tawil, D., & Velonaki, M. (2015). Group vs. individual comfort when a robot approaches. *Social Robotics*, LNCS 9388, 41–50.

Chang, W. L., & Šabanović, S. (2015, March). Interaction expands function: Social shaping of the therapeutic robot PARO in a nursing home. In *Proceedings of the tenth annual ACM/IEEE international conference on human-robot interaction* (pp. 343–350).

Clarke, A. E., Friese, C., & Washburn, R. S. (2017). *Situational analysis: Grounded theory after the interpretive turn.* Sage publications.

Coronado, Enrique, et al. "Visual programming environments for end-user development of intelligent and social robots, a systematic review." *Journal of Computer Languages* 58 (2020): 100970.

Crabtree, A., Tolmie, P., & Chamberlain, A. (2020). "Research in the Wild": Approaches to understanding the unremarkable as a resource for design. *Into the wild: Beyond the design research lab* (pp. 31–53). Springer.

Fang, Bin, et al. "Survey of imitation learning for robotic manipulation." *International Journal of Intelligent Robotics and Applications* 3 (2019): 362–369.

Friedman, B., Kahn, P., & Borning, A. (2002). Value sensitive design: Theory and methods. *University of Washington Technical Report*, *2*(8), (pp. 1–8).

Garfinkel H. (1967). *Studies in ethnomethodology*. Prentice Hall.

Gemeinboeck, P. (2021). The aesthetics of encounter: a relational-performative design approach to human-robot interaction. *Frontiers in Robotics and AI*, *7*, 577900.

Gonzalez, I. (2022). Robot code of conduct for automated dairy farming. http://resolver.tudelft.nl/uuid:3ae4c065-bbbc-4472-9915-739f629ea440

Hagenow, Michael, et al. "Corrective shared autonomy for addressing task variability." *IEEE Robotics and Automation Letters* 6.2 (2021): 3720–3727.

Kelley, S. W., Hoffman, K. D., & Davis, M. A. (1993). A typology of retail failures and recoveries. *Journal of Retailing*, *69*(4), 429–452.

Lassiter, L. E. (2005). *The Chicago guide to collaborative ethnography*. University of Chicago Press.

Latour, B. (2007). *Reassembling the social: An introduction to actor-network-theory*. Oup Oxford.

Lee, H. R., & Riek, L. (2023). Designing robots for aging: Wisdom as a critical lens. *ACM Transactions on Human-Robot Interaction*, *12*(1), 1–21.

Lee, H. R., & Riek, L. D. (2018). Reframing assistive robots to promote successful aging. *ACM Transactions on Human-Robot Interaction (THRI)*, *7*(1), 1–23.

Lee, H. R., Šabanović, S., & Kwak, S. S. (2017, May). Collaborative map making: A reflexive method for understanding matters of concern in design research. In *Proceedings of the 2017 CHI conference on human factors in computing systems* (pp. 5678–5689).

Lee, H. R., Tan, H., & Šabanović, S. (2016, August). That robot is not for me: Addressing stereotypes of aging in assistive robot design. In *2016 25th IEEE international symposium on robot and human interactive communication (RO-MAN)* (pp. 312–317). IEEE.

Lee, H. R., Tan, X., Zhang, W., Deng, Y., & Liu, Y. (2023, August). Situating robots in the organizational dynamics of the gas energy industry: A collaborative design study. In *2023 32nd IEEE international conference on robot and human interactive communication (RO-MAN)* (pp. 1096–1101). IEEE.

Lee, M. K., Kiesler, S., Forlizzi, J., Srinivasa, S., & Rybski, P. (2010, March). Gracefully mitigating breakdowns in robotic services. In *2010 5th ACM/IEEE international conference on human-robot interaction (HRI)* (pp. 203–210). IEEE.

Leonardi, Nicola, et al. "Trigger-action programming for personalising humanoid robot behaviour." *Proceedings of the 2019 CHI Conference on Human Factors in Computing Systems*. 2019.

Mindell, D. (2015). *Our robots, our selves: Robotics and the myth of autonomy*. Viking Press.

Moesgaard, F., Hulgaard, L., & Bødker, M. (2022, August). Incidental encounters with robots. In *2022 31st IEEE international conference on robot and human interactive communication (RO-MAN)* (pp. 377–384). IEEE.

Murray-Rust, D., Lupetti, M. L., Ianniello, A., Gorbet, M., Van Der Helm, A., Filthaut, L., … & Beesley, P. (2024, March). Spatial robotic experiences as a ground for future HRI speculations. In *Companion of the 2024 ACM/IEEE international conference on human-robot interaction* (pp. 57–70).

Mutlu, B., & Forlizzi, J. (2008, March). Robots in organizations: The role of workflow, social, and environmental factors in human-robot interaction. In *Proceedings of the 3rd ACM/IEEE international conference on Human robot interaction* (pp. 287–294).

Ocnarescu, I., & Cossin, I. (2017). Rethinking the why of socially assistive robotics through design. In *Social robotics: 9th international conference, ICSR 2017*, Tsukuba, Japan, November 22–24, 2017, Proceedings 9 (pp. 383–393). Springer International Publishing.

Ocnarescu, I., & Cossin, I. (2020). Discovery report following 5 years of research project on socially assistive robotics. *Current Robotics Reports*, *1*, 269–278.

Paxton, Chris, et al. "CoSTAR: Instructing collaborative robots with behavior trees and vision." *2017 IEEE international conference on robotics and automation (ICRA)*. IEEE, 2017.

Pot, Emmanuel, et al. "Choregraphe: a graphical tool for humanoid robot programming." *RO-MAN 2009-The 18th IEEE International Symposium on Robot and Human Interactive Communication*. IEEE, 2009.

Reig, S., Luria, M., Wang, J. Z., Oltman, D., Carter, E. J., Steinfeld, A., Forlizzi, J., & Zimmerman, J. (2020, March). Not some random agent: Multi-person interaction with a personalizing service robot. In *Proceedings of the 2020 ACM/IEEE international conference on human-robot interaction* (pp. 289–297).

Robinson, F. A., Velonaki, M., & Bown, O. (2021). Smooth operator: Tuning robot perception through artificial movement sound. In *Proceedings of the 2021 ACM/IEEE international conference on human-robot interaction*, 2021.

Rosenthal-von Der Pütten, A., Sirkin, D., Abrams, A., & Platte, L. (2020, March). The forgotten in HRI: Incidental encounters with robots in public spaces. In *Companion of the 2020 ACM/IEEE international conference on human-robot interaction* (pp. 656–657).

Rossano, G. F., Martinez, C., Hedelind, M., Murphy, S., & Fuhlbrigge, T. A. (2013, August). Easy robot programming concepts: An industrial perspective. In *2013 IEEE international conference on automation science and engineering (CASE)* (pp. 1119–1126). IEEE.

Šabanović, S. (2010). Robots in society, society in robots: Mutual shaping of society and technology as a framework for social robot design. *International Journal of Social Robotics*, *2*(4), 439–450.

Savoia, A. (2011). Pretotype it. *Make sure you are building the right it before you build it right*. Retrieved online on July 31st, 2024, at: https://www.pretotyping.org/uploads/1/4/0/9/14099067/pretotype_it_2nd_pretotype_edition-2.pdf

Senft, Emmanuel, et al. "Situated live programming for human-robot collaboration." *The 34th Annual ACM Symposium on User Interface Software and Technology*. 2021.

Shin, J. Y., Chaar, D., Davis, C., Choi, S. W., & Lee, H. R. (2021). Every cloud has a silver lining: Exploring experiential knowledge and assets of family caregivers. *Proceedings of the ACM on Human-Computer Interaction*, *5* (CSCW2), 1–25.

Silvera-Tawil, D., Rye, D., & Velonaki, M. (2014). Interpretation of social touch on an artificial arm covered with an EIT-based sensitive skin. *International Journal of Social Robotics*, *6*(4), 489–505.

Silvera-Tawil, D., Velonaki, M., & Rye, D. (2015, August). Human-robot interaction with humanoid Diamandini using an open experimentation method. In *2015 24th IEEE international symposium on robot and human interactive communication (RO-MAN)* (pp. 425–430). IEEE.

Smith, A. K., Bolton, R. N., & Wagner, J. (1999). A model of customer satisfaction with service encounters involving failure and recovery. *Journal of Marketing Research, 36*(3), 356–372.

Stegner, L., Senft, E., & Mutlu, B. (2023, April). Situated participatory design: A method for in situ design of robotic interaction with older adults. In *Proceedings of the 2023 CHI conference on human factors in computing systems* (pp. 1–15).

Steinmetz, Franz, Annika Wollschläger, and Roman Weitschat. "Razer—a hri for visual task-level programming and intuitive skill parameterization." *IEEE Robotics and Automation Letters* 3.3 (2018): 1362–1369.

Suchman, L. A. (2007). *Human-machine reconfigurations: Plans and situated actions.* Cambridge University Press.

Turkle, S. (2011). *Along together: Why we expect more from technology and less from each other.* Basic Books

Umbrello, S., & Van de Poel, I. (2021). Mapping value sensitive design onto AI for social good principles. *AI and Ethics, 1*(3), 283–296.

Van Wynsberghe, A. (2013). Designing robots for care: Care centered value-sensitive design. *Science and Engineering Ethics, 19*, 407–433. https://doi.org/10.1007/s11948-011-9343-6

Velonaki, M., Scheding, S., Rye, D., & Durrant-Whyte, H. (2008). Shared spaces: Media art, computing, and robotics. *Computers in Entertainment (CIE), 6*(4), 1–12.

Wright, J.A. (2023). *Robots won't save Japan: An ethnography of eldercare automation.* ILR Press.

4 Designing Human–Robot Ecologies

Beyond Utilitarian Relations

Nazli Cila
Delft University of Technology, Delft, Netherlands

4.1 INTRODUCTION

Picture a factory where sophisticated robotic coworkers are introduced. Initially designed to streamline production processes, these robots inevitably influence the way human workers perceive and engage with their tasks. As humans collaborate with robots, their understanding of labor, efficiency, and even craftsmanship may evolve, leading to a redefinition of work practices and professional identities within the factory ecosystem.

This transformation is not confined to factory floors alone. Whether it be a care home, a bustling public space, a dynamic classroom, or the sanctity of one's own home, the introduction of a robot ignites complex shifts in everyday practices, transcending utilitarian relations. Robots are not standalone entities; they become enmeshed within existing and constantly evolving ecosystems comprised of people, things, and other nonhumans, and social constructs. The interactions between humans and robots cease to be mere transactions and instead evolve into a continuous negotiation within these complex webs of relationships.

As humans, our existence is intricately interwoven with the material world that surrounds us. While our tool use, epitomized by Heidegger's iconic hammer, has long captivated our understanding of the human–technology relationship, it is in the contemporary age of pervasive smart technologies that this entanglement between humans and technologies takes on new dimensions (Frauenberger, 2021). This paradigm shift has garnered attention from scholars in Human–Computer Interaction (HCI), including Giaccardi and Redström (2020), Frauenberger (2019), Wakkary (2020), Rozendaal et al. (2019) among others, who draw upon relational ontologies and entanglement theories from the humanities to construct a nuanced picture of the dynamic interplay between humans and technologies while acknowledging their inherent interrelatedness.

DOI: 10.1201/9781003371021-4
This chapter has been made available under a CC-BY-NC-ND license.

Postphenomenology, notably, has served primarily as an analytical framework to elucidate these relationships, compelling us to reassess our perceptions and interactions with robots. Expanding upon Ihde's (1990) seminal work, Verbeek (2015) delineated four categories of human–technology relations. Embodiment relations involve the augmentation of bodily perception, where technology becomes an extension of ourselves (e.g., wearing glasses). Hermeneutic relations highlight technology's role as an intermediary that aids in interpreting the world (e.g., thermometer, microscope). Background relations manifest when technologies subtly blend into the environment, shaping our experiences without notice (e.g., heating system). The fourth category, alterity relations, as described by Coeckelbergh (2011), pertains to human–robot interactions (HRI), viewing the robot as an "other." Alterity relations involve relating to the robot as a distinct entity. Correspondingly, Hassenzahl et al. (2020) coined the term "otherware" to denote such proactive, self-learning, artificial intelligence (AI)-infused systems that humans interact with through cooperation, delegation, commands, and trust. Their call-to-action urges designers to develop interaction paradigms for otherware, including robots, that embody desirable qualities of interaction.

The methods and approaches outlined in this section offer a means to unpack these alterity relations by identifying, assessing, imagining, and reimagining them, all with the overarching goal of fostering a broader design paradigm. Many of these approaches draw inspiration from entanglement theories (e.g., postphenomenology, activity theory, posthumanism), aiming to reveal and engage with the socio-material ecologies that robots are or become part of. The approaches acknowledge the dynamic and multifaceted nature of these ecologies, recognizing that they evolve over time amidst different social and political contexts, subjected to numerous influences and stresses. Consequently, they prioritize the situatedness and contextuality of HRI. Moreover, by investigating the alterity of robots and the various experiences they evoke, many of these approaches offer valuable handles on accountability and the ethical dimensions inherent within human–robot ecologies.

The first two approaches presented in this section are performative in nature, offering methods that facilitate a shift in one's position within a human–robot relation, enabling people to adopt the perspective of the "other." These methods prioritize "decentering the human" (Nicenboim, et al., 2023), encouraging designers to critically reflect on their biases, positions, and inherent limitations.

In Box 4.1, Nicenboim and Giaccardi introduce *"Conversations with Agents,"* a technique designed to reveal overlooked human and nonhuman viewpoints within a design space. Employed during the ideation phase, this technique assists in framing the design challenge, thereby potentially unveiling nuanced aspects of the design space that might not have been initially contemplated. When utilized during the evaluation phase, it has the capacity to challenge existing preconceptions and conceptualizations associated with robots.

In Box 4.2, Dörrenbächer presents *"Techno-Mimesis"* in which humans assume the role of robots using props, thereby gaining insight into the world from a robotic perspective. The primary objective of Techno-Mimesis is to enable designers to simultaneously experience the human and robotic condition, highlighting their differences and facilitating the identification of potential robotic capabilities in terms of physical, cognitive, and communicational attributes.

In a postphenomenological understanding of human–robot relations, agency is not seen as an inherent attribute of either humans or robots, but rather as something that emerges through their interactions (Coeckelbergh, 2011). Thus, fundamentally, any product can exhibit agency based on how it is experienced. However, the intelligence inherent in robots endows them with a unique capability to express intent in ways that traditional, non-intelligent objects cannot. The intent imbued within a robot can be both more implicit and more impactful than that which could embed within a tool like a hammer (Frauenberger, 2021). Consequently, robots possess a stronger claim to agency, and their potential to shape, guide, or manipulate human activity is significantly amplified. The following three methods in this section delve into the agency and intentions of robots, exploring how these elements can be designed and communicated with people in responsible, engaging, and effective ways.

In Box 4.3, Rozendaal initiates this exploration by ensuring that the agency exhibited by robots is fitting and meaningful within the context in which they are introduced. The *"Intentionality"* framework aims to assist designers in carefully considering the framing, embedding, and transformation of robots within specific contexts, thereby designing robots whose form and expressivity align with their purpose and environment.

In Box 4.4, Laschke and Hassenzahl introduce the *"Aesthetic of Friction"* as an approach for leveraging the agency of smart technologies to guide people toward more responsible choices, such as those concerning their well-being, collective welfare, or environmental impact. Here, agentic technologies express their intentions by serving as "pleasurable troublemakers," prompting individuals to confront unhealthy behaviors while offering pathways toward more desirable alternatives.

In Box 4.5, Okada proposes *"Symbiosis,"* which is a perspective that embraces the incompleteness and potential shortcomings of robots, using these characteristics to solicit assistance from people. These human-dependent social robots actively engage individuals to achieve shared goals, fostering a harmonious coexistence between robots and humans.

Designing human–robot ecologies necessitates a shift in perspective for designers, moving away from viewing the design task solely in terms of the individual actors involved, and instead considering the collective outcomes of their entanglement. How, then, should designers approach this newly expanded design space? The designer needs to assume a less authoritative stance, acknowledging that the robot evolves into an entity in its own right, susceptible to appropriation by humans in the process of making sense of their environment. It is the designer's role to shape the robot, configuring its technology, appearance, and behavior. These configurations facilitate different relations, thus requiring design approaches that emphasize openness and ambiguity (Van Beek et al., 2023). The following two approaches embody this by serving as means for creating open-ended robots that people can interpret and interact with in various ways, while also encouraging reflection.

In Box 4.6, Hoggenmueller's *"Playfulness"* approach endeavors to design interactions that stimulate imagination and creativity, transcending the traditional focus on "utility" in robot design. The ludic design principles elicited from this approach aim at designing robots that entice curiosity and invite exploration.

In Box 4.7, Pierce introduces *"Para-functionality,"* an approach that extends beyond conventional definitions of functionalism to incorporate poetic elements into robot design. The open-ended nature of the robots enables people to generate diverse interpretations regarding the robots' form, intentions, behaviors, and ultimately unveiling social tensions and ethical challenges.

This section of the book advocates for HRI research and robot design practices to view relations with robots as integral components of a broader relationship between humans and their environment, mediated by technology, of which robots are a constituent part.

In Box 4.8, Murray-Rust introduces *"Relationality"* and encourages us to perceive the robot not as an isolated artifact, but rather to examine it within the context of surrounding practices. Much like humans are not confined solely to biology but are also influenced by culture, society, and values, robots are embedded within complex socio-technical ecosystems.

As I conclude this introduction to the forthcoming methods, let's revisit the insights of Verbeek (2015) who aptly expressed, "What is being designed, then, is not a thing but a human–world relation in which practices and experiences take shape" (p. 26). Our aspiration is for the HRI community to embrace this holistic perspective, viewing robots not as standalone artifacts but as integral parts of socio-technical ecologies. This perspective prompts a deeper reflection on their broader role and impact within society, extending beyond mere utility.

BOX 4.1 CONVERSATIONS WITH AGENTS

Iohanna Nicenboim and Elisa Giaccardi

Conversations with Agents supports designers and researchers in exploring and reimagining intelligent agents such as conversational interfaces (e.g., AI-powered artifacts and robots) by taking a **more-than-human perspective**. It integrates several design techniques, including interviews, enactments (role-playing), and speculation. More precisely, it builds upon the technique known as Interview with Things (Reddy et al., 2021) and expands it toward more performative speculations.

When used in the ideation phases of a design process, this technique can help to frame (or reframe) the design challenge, by revealing more complex and nuanced aspects of the design space that were perhaps not initially considered. When used for evaluating the implications of existing technologies, it can help to understand and question the existing imaginaries associated with a particular product, or the harmful biases that might have been embedded in its interaction design. For example, in research on conversational interfaces, this technique has illustrated how the interactions of devices such as Alexa and Google Home exacerbate gender, racial, and anthropocentric biases (Nicenboim et al., 2020); and how those stereotypes can be particularly harmful when coupled with a lack of positionality—when the agent's knowledge, ownership, and biases are not accounted for or explained in its interaction. Later, the technique has been proven to be useful in ideating alternative interactions that are more situated and inclusive (Nicenboim et al., 2023; Reddy et al., 2021).

Conversations with Agents situates the imaginaries and shortcomings of existing technologies within the vast infrastructures that underlie the complex socio-technical systems in which current agentic technologies are embedded. As **a decentering technique**, it not only reveals overlooked human and nonhuman perspectives that are useful in the design of intelligent agents; it also defamiliarizes the designer's perspective, inviting them to be accountable for their position and limitations, such as the social and political context that shapes the designer's view of the world, and how that influences and biases their research. In doing so, it provides more than a critique to existing interaction designs—it actively suggests alternatives. It proposes different interactions the agent might have, it points at perspectives that can be included, and actively explores new affirmative relations based on values such as inclusion, care, and reciprocity (Nicenboim et al., 2020; Reddy et al., 2021).

FIGURE 4.1 The second activity of the workshop was to interview a robot. To do that, one participant asked questions, another participant enacted the robot's responses with voice and movements, and a third participant took notes.

WORKSHOP "IN CONVERSATION WITH ROBOTS"

Conversations with Agents were used in several workshops. At the conference Designing Interactive Systems (DIS) in 2020 and the Mozilla Festival in 2021, we worked with conversational agents; at the Research Through Design Conference (RTD2019) and in several classes at the Delft University of Technology between 2020 and 2023, we worked with intelligent agents more generally; and at the conference Thingscon in 2022, we worked with robots.

In the workshop "In Conversation with Robots," human participants were first invited to interact with different devices: Cosmo (a toy robot), a vacuum cleaner, and an Amazon Alexa. Then participants conducted interviews with the robots and enacted their responses with voice and/or movements by using speculation and role-play tactics. Based on the insights that emerged from the interviews, participants prototyped alternative interactions with the robots.

The prototypes challenged the pervasive narratives of efficiency and gender stereotypes that often accompany the design and development of robots. For example, it unveiled assumptions of "cleaning" as an activity that is easy to do and thus easy to replace, and the imaginary of "the home" as a flat and enclosed space. It also helped them contest the anthropocentric notion that robots need to be functional tools for people by highlighting autonomous interactions that do not involve humans at all.

Furthermore, the technique unsettled traditional imaginaries of gender, efficiency, and automation by enabling emancipated interactions and by reconfiguring agency and control in HRI. To contest the existing narratives with alternative ones, the workshop's outcomes expanded the design space to consider failures and misunderstandings as potential areas in which designers could make the robot's limitations and infrastructures visible, supporting more responsible and explainable interactions.

A key takeaway of the workshop was that along with designing interactions with robots that are more efficient, it is of paramount importance that the interaction is able to communicate the robot's limitations—to allow people to develop their own sense of trust—and to reveal the socio-technical infrastructures in which the robots are embedded, to get a sense of what are the broader implications of interacting with them beyond everyday contexts.

BOX 4.2 TECHNO-MIMESIS

Judith Dörrenbächer

Techno-Mimesis (Dörrenbächer et al., 2020) is a performative method. It aims to rethink existing technology. When applied to social robotics, it enables to emphasize "robotic superpowers" that distinguish robots from humans. The method is all about a bodily transformation: Humans temporarily turn into robots and **perceive a use scenario from the robot's perspective**. To enable moving and sensing in technologically determined ways, so-called prostheses are applied. These might be low-tech mockups, simply made from cardboard. Typical input and output modalities (e.g., speech recognition) and well-known hardware (e.g., a platform with wheels) serve as rationales. An example is eyeglasses that change the human visual sense to a constrained or enhanced vision.

Of course, these mockups do not copy robotic sensing and movement perfectly. However, they allow to change perspective and get an embodied understanding of robots. Right after the transformation, the "robot" acts out a use scenario with other humans who act as users (e.g., seniors in a care facility). Subsequently, the "robot" and the "users" undergo an interview and answer questions about their experiences (e.g., "In which situations during the scenario did you feel positive about being a robot?"). Compared to the Wizard-of-Oz technique, Techno-Mimesis does not aim at improving usability, but at experiencing being human and being robot at the same time, making their differences apparent and facilitating the discovery of robotic strengths, such as always being patient or being objective. I derived Techno-Mimesis from the animistic practice of mimesis, which was defined by the anthropologist Rane Willerslev (2007). Similar to indigenous people, who practice mimesis to control the relationship to other species, designers practice mimesis to control the relationship to their own creations. By slipping into the corpus of their robot, designers are able to experience unique robotic abilities vis-à-vis their own unique humanness. This allows to make use of the full potential of robots' social abilities instead of just imitating humans (anthropomorphism).

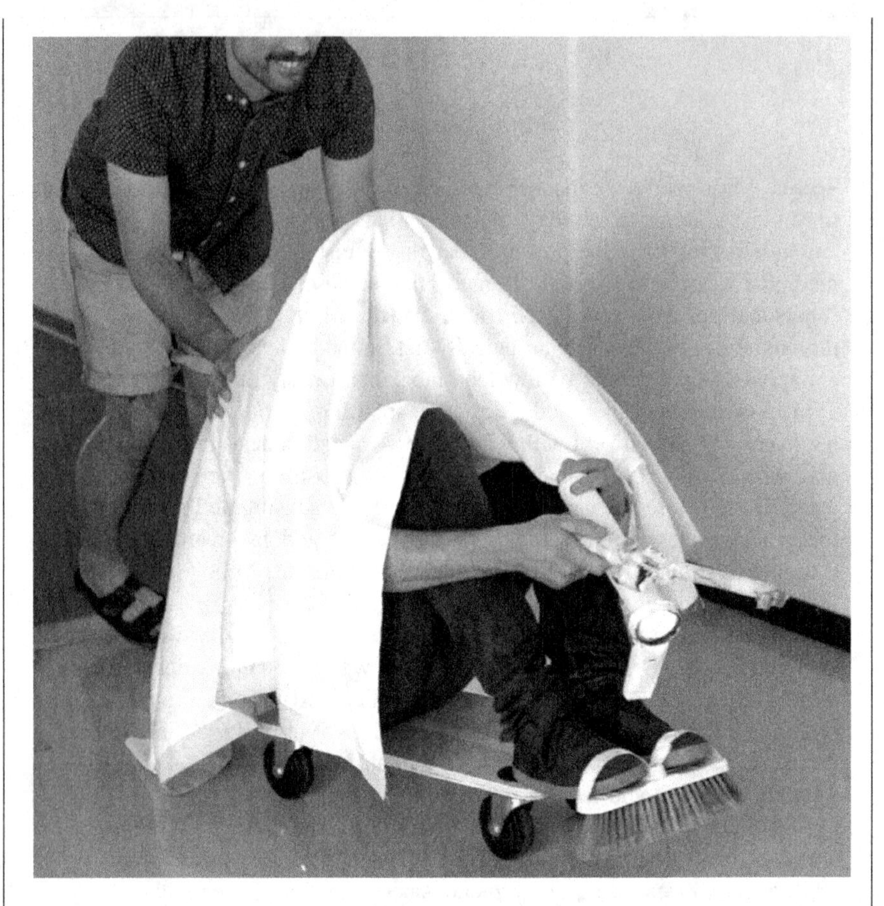

FIGURE 4.2 Being human and becoming a cleaning robot for train stations at the same time. A robot designer playacting a scenario with the help of our perception changing prostheses.

TECHNO-MIMESIS WITH A SHOPPING ROBOT

Together with my colleagues Marc Hassenzahl and Diana Löffler, I applied Techno-Mimesis in robotics with three different service robot development projects. The first team developed a shopping robot for supermarkets; the second one created two cleaning robots (domestic and public setting); and the third team developed a robot to assist visually impaired people in handing over objects.

Together with colleagues of the University of Siegen, we familiarized ourselves with the hardware and software the teams used for their robots. We prepared prostheses accordingly, for example, a basket for the shopping robot, a torch for the cleaning robot (since it was supposed to project signals onto the floor), a stick that vibrated when obstacles appeared on its way (simulating ultrasonic sensors).

We met the teams individually and asked one of the designers to transform his or her body. For example, the designer becoming the shopping robot sat on a dolly board and tied one of his arms to his back to simulate being equipped with one gripper only. Further, he wore glasses on the back of his head to simulate a 360° view and he wore a tablet tied to his neck to allow communication with customers of the supermarket. The remaining designers slipped into the roles of users, and all of them playacted use scenarios, e.g., doing cooperative shopping.

The interviews that followed the role-play revealed a series of useful "robotic superpowers." For example, the "robots" experienced the advantage of being unambiguous and straightforward when approaching pedestrians. Being unempathetic prevented collisions and misunderstandings. If the robot made use of his voice output, i.e., behaving human-like, confusion or even embarrassment resulted. Customers of the supermarket, for example, loved the robot's neutrality and thus made it shop condoms or mountains of chocolate. They did not want to use human language to express their wishes in public, and they were happy not having to say thank you to the "robot." Further, the designers of the cleaning robot for public spaces realized it to be an advantage not being able to identify gender, sex, or race of people. This way, they were not able to discriminate.

In addition, the "robot" for visually impaired people realized to be supernaturally patient, what allowed users to repeat their questions endlessly.

Most superpowers were revealed because of the double perspective of being human and robot at the same time: The participating "robots" spontaneously responded in a human way to the scenarios—they felt pushed around, ashamed, or indiscreet—feelings real robots certainly do not have. However, with the help of our perspective-changing prostheses, they realized: being a robot could actually be an advantage for the scenario. Following the experiences of Techno-Mimesis, our robot designers explained they might rethink anthropomorphic design, such as the usage of voice recognition or the conventions of politeness, and to make use of the robot's unique superpowers instead, such as its thing-like simplicity or neutrality. We realized, in social robotics the approach draws attention to alternatives to anthropomorphism, what is still the dominating design strategy.

BOX 4.3 OBJECTS WITH INTENT

Marco C. Rozendaal

Objects with Intent (OwI) is a conceptual framework that helps to shape robots as intelligent everyday artifacts. The framework draws from Activity Theory (AT; Kaptelinin & Nardi, 2006; Leontiev, 1975) and Dennett's (1989) theory of intentionality. AT emphasizes the role of motives in human activity and considers the physical and social contexts in which people operate. AT further emphasizes how activities can be analyzed in terms of the actions and operations that are required to fulfill a motive. Tools play a significant role as mediators of human activity and have cultural significance due to their historical development. Dennett's theory of intentionality extends intentionality to all entities, not just living beings, attributing beliefs and desires to explain their rational actions. The OwI framework therefore merges a cultural–historical perspective on artifacts with an intentional stance to explain their intelligent behavior (Rozendaal, 2016; Rozendaal et al., 2019; Rozendaal et al., 2020). The framework encompasses three key aspects for designers to consider:

Framing: Crafting robots as hybrid artifacts that integrate cultural expression with intelligent behavior. This deliberate hybridity fosters interactions that appeal to curiosity and blends instrumental with socially communicative interactions.

Embedding: Recognizing how interactions with robots are enmeshed and coordinated with other things, people, and spaces. This understanding informs a contextual approach to interaction as they are experienced and informs an object's sensing and actuation capabilities.

Transformation: Acknowledging the co-development process between humans and robots, where each learns from the other. Interactions with robots transition through different phases based on individual and social factors, informing how a robot's behavioral repertoire can be designed to align with this dynamic.

The OwI framework serves as a guide to envision robots as smart everyday objects and supports the crafting of models and prototypes when used in an evaluative way to better understand how they perform in particular contexts of use.

A ROBOT BALL FOR PHYSICAL PLAY

A robot ball concept was created to mobilize children in their patient rooms while being hospitalized. The OwI framework was used to inform the conceptualization and evaluation of the robot ball as an everyday artifact with intelligent behavior. A prototype was built and tested in patient rooms using a Wizard-of-Oz technique in which the researcher enacted the ball's intelligence. In this method, the researcher controlled and animated the robot ball based on the unfolding interactions within certain rules that intended to trigger physical activity yet allowing room for improvisation.

We continue by exploring how the method's key concepts of framing, embedding, and transformation were used to analyze the results and what insights they provided.

"A robot as an everyday ball with intelligent behavior." The framing concept guided the exploration of what kind of artifact with expressive intelligent

FIGURE 4.3 Impression of Fizzy rolling around.

behavior could stimulate physical activity. A ball was chosen as an archetypal artifact that is associated with physical, free, and social play. The ball was designed with a material embodiment that allowed it to be robust when used for physical play. By endowing the ball with intent and intelligent behavior, it enhanced its capabilities beyond a typical ball, fostering flexibility in play by augmenting existing games, enhancing exploration, and triggering social play. Its unique hybrid character offered advantages in interaction compared to robots that rely on explicitly human schemas.

"Observing human–robot interactions played out in context." Children's interactions with the ball are enmeshed with mundane activities, such as eating, socializing, and routine tasks performed by nurses. Children's play was often shaped by parental suggestions and could lead to play activities that also involved the parents. The physical space became part of these interactions too. For example, playing hide-and-seek involved the ball moving behind furniture, and using the robot ball in a game of bowling required making an isle and gathering cups as targets. These observations underscore the importance of designing the intelligent behavior of robots with consideration for context, both in terms of its limitations and opportunities.

"Encountering robots and sustaining interaction." The introduction of the robot ball into the patient room changed the atmosphere into a more playful one. The initial encounter with the robot ball led to exploring the ball's capabilities, after which stable interaction patterns were established. These patterns varied given the child's interests and onset of boredom or fatigue. Interactions were further influenced by social factors. Parents suggested games that children continued to play alone, or children invited parents to participate. These observations demonstrate that meanings attributed to robots evolve through exploration, resulting in stable usage patterns that vary based on individual and social influences. To encourage sustained interactions with robots, it is essential for their behavior to adapt to these changing dynamics.

Designing robots with the OwI framework enables the creation of robots that transcend common robot stereotypes. This approach ensures that their form and expressivity align with their purpose and context. It also enables that their behavior repertoire is sensitive to the dynamic and evolving relationships established within their environment.

BOX 4.4 AESTHETICS OF FRICTION

Matthias Laschke and Marc Hassenzahl

Driving less, eating less, and reducing heating are sensible resolutions for behavior change, but implementing them in everyday life can be challenging. One of the main obstacles are everyday routines, which often lead to mindless, automatic actions. This tendency is also reflected in the design of interactive technologies, which focus on convenience over promoting change. For example, an assistive shopping robot that suggests products based on past behavior may enhance convenience but cannot facilitate change.

To foster behavior change, technology should disrupt routines and help shape alternative practices aligned with individual goals. This requires introducing a **controlled level of friction** that remains acceptable. We suggest Aesthetic of Friction (AoF; Hassenzahl & Laschke, 2014) as an approach to designing interactive technologies that actively disrupt routines the moment they occur. Agentive chairs and office can disrupt sitting habits, shopping robots can disrupt buying habits, or advanced driver-assistance can disrupt driving habits. Friction, by definition, needs to be introduced actively by the object itself. Consequently, these objects are distinct from everyday tools. They are not readily available extensions of ourselves—they do not enhance comfort. Instead, they have their own agenda, becoming counterparts or "otherware" (Hassenzahl et al., 2020; Laschke et al., 2020).

Friction, however, can be unnerving, and technologies that create friction risk being rejected as unwanted troublemakers. To address this, friction should be coupled with acceptance. For instance, a troublemaking shopping robot should not only disrupt the purchase of sugary drinks but also offer healthier alternatives. The friction highlights unconscious behavior, while the alternative provides a pathway toward better choices. Troublemakers should also understand and accommodate reasons for deviating from the suggested behavior, allowing conscious disregard of recommendations. This can be achieved through flexibility, humor, and recognizing the complexity of everyday decision-making.

By combining alternatives and understanding, friction can become more bearable and even pleasurable. Troublemakers can **facilitate behavior change** without imposing their agenda, becoming what is referred to as "Pleasurable Troublemakers." These interactive technologies aim to increase the likelihood of behavior change while respecting individual choices and constraints.

FIGURE 4.4 The Keymoment with a bike and car key side by side creating a choice architecture.

KEYMOMENT THAT NUDGED ME TO (RE-)THINK

Keymoment (Laschke et al., 2014) tackles the challenge of both contributing to the environment and promoting personal physical activity. The World Health Organization (2010) suggests incorporating cycling into daily life, such as commuting to work, to address these objectives. However, implementing this objective is easier said than done. Psychological approaches offer strategies to facilitate this process. For instance, following Gollwitzer's (1999) concept of Implementation Intention, which essentially connects specific situations with goal-oriented behaviors, one should consistently question whether it is possible to opt for cycling whenever there is a need to go somewhere. According to Gollwitzer, such a "simple plan" helps to use the bike (more frequently) and achieve one's goal in the long run.

Keymoment serves as a materialized Implementation Intention. It is a key rack placed in a common location, such as the hallway or by the front door. The rack holds the bike key and the car key side by side. By doing so, Keymoment confronts people with a choice between biking or driving. If you take the bike key, you have chosen the alternative that leads to your "desired" goal. However, choosing the car key, consciously or unconsciously, the bike key falls to the floor, which most people then pick up. Holding both keys in your hands, you face a genuine dilemma—do you take the bike or stick with the car? In this manner, Keymoment interrupts the habit of reaching for the car key and prompts reflection. Hence, Keymoment is not just a simple wooden box but rather a kind of autonomous roommate, just waiting to intervene. Additionally, Keymoment offers a goal-oriented alternative in the form of the bike.

Nevertheless, the act of the bike key falling can be bothersome. After all, you intended to go by car rather than by bike. Suddenly, you find yourself in a situation where you must justify your behavior. Furthermore, you are now required to engage in thought (in other words, reflect) about something that you previously did effortlessly without much reflection (habitually).

Like an agentive assistant or tiny robot, Keymoment nudges individuals to make a reflective decision and presents them with a choice. To make the friction created by Keymoment more bearable, the AoF includes some design strategies. For example, you can take a break by placing your bike key on top of the Keymoment. Also, cheating the system is part of its design. You can exchange the bike key with the car key, resulting in Keymoment suggesting the car key even if the person initially reached for the bike key. However, this brief moment of contentment is short-lived, as it is difficult to cheat oneself—an ironic twist. Other elements of understanding could also be considered. For example, Keymoment could suspend the bike key's falling when it rains.

Keymoment is a highly conceptual piece that exemplifies the fine line between friction and understanding. Friction, from our perspective, is a vital element in driving transformation. At the same time, understanding the difficulty of changing is crucial in perceiving transformation as a process, an evolution of people. Such a design needs a normative standpoint alongside a good portion of irony, humor, and empathy.

BOX 4.5 SYMBIOSIS

Michio Okada

We have fostered a culture that values self-reliance as a virtue. Extending this mindset to even how we raise our children, we encourage them to learn to conquer simple tasks like putting on their socks independently at an early age, and they may proudly exclaim, in response, "Look! I did it all by myself! Isn't it awesome?"

A similar inclination is observed in the field of autonomous robotics. Historically, the pursuit of autonomous robots has revolved around the individual capability-based approach, striving to embed all the desired functions and performance within a single robot. This design perspective, known as "design by addition" or the "Swiss Army Knife approach," entails a continuous process of relentlessly adding a multitude of functions to maximize the robot's capabilities. Yet, even the most advanced robots and AI technologies inevitably possess many flaws and shortcomings. Instead of boasting about all the things that they can do, why not empower robots to recognize their weak points and seek help from those around them for tasks that surpass their capabilities?

Drawing from such thoughts, a concept has emerged: "human-dependent social robot" or **a weak robot**—a robot that is intentionally incomplete on its own. Instead, it embraces a half-receptive nature toward others, skillfully engaging them to work together to achieve a shared goal. Weak robots, with their **charming imperfection**, evoke a sense of cuteness that naturally compels us to lend a helping hand.

Our symbiotic design (Okada, 2022) approach centers around weak robots and explores how we can design and foster relationships between humans and robots based on recognizing robots as inherently imperfect. Through this approach, we aim to cultivate harmonious coexistence, where the strengths and weaknesses of both humans and robots are mutually enhanced and supported (https://www.icd.cs.tut.ac.jp).

"WEAK ROBOTS" FOR SYMBIOTIC RELATIONS WITH HUMAN

We hold the hope that useful robots will enhance and enrich our lives. Yet, what would life be like if we were surrounded only by such convenient robots? Is there any room left for our own active participation? What if a robot demanded a bit more engagement from us? These contemplations birthed the concept of the Sociable Trash Box, which cannot independently pick up trash and elicits children's help in this endeavor. In contrast to an action strategy based on

FIGURE 4.5 iBones handing out pocket tissues.

individual competencies, this approach is called a relational-oriented action —It involves individuals successfully achieving their objectives by actively seeking and eliciting help from social others.

With this vision in mind, we developed and introduced the Sociable Trash Box (Yamaji et al., 2011) to a playground where children were playing. The peculiar appearance of the trash box immediately caught their attention as they gathered around the robot and asked: "What is it?" Perhaps getting a sense of the robot, a child took the paper bag in their hand and tossed it into the Sociable Trash Box. The robot responded with a bow-like gesture, leaving it uncertain whether it was an expression of gratitude for the assistance or a plea for further help. Encouraged by this gesture, the nearby children joined in, enthusiastically searching for more garbage, eventually filling the robot's storage space. This somewhat unreliable robot, which cannot pick up trash by itself, successfully achieved its goal of collecting trash by engaging and interacting with the children.

Another example of a weak robot is our iBones robot, which endeavors to distribute pocket tissues to passersby on street corners, employing a common guerrilla marketing method in Japan. With the iBones robot, pocket tissues are offered to strangers strolling by, with the robot being alien to many. The successful exchange of the tissue relies on both parties: If the person declines to accept, the tissue cannot be handed out; likewise, if the robot fails to deliver the tissue properly, it cannot be received. Both parties must partially trust the other in this action, and the exchange fails if the shared goal is absent.

...Whenever a person approaches the robot, it attempts to hand out the tissue. If it predicts that the action will fail, the robot hesitates and repeats the motion, displaying a somewhat fidgety behavior. Perhaps feeling bad for the robot, an older woman stopped before the robot and happily accepted a tissue, synchronizing her movements with the iBones robot's hand. Her happiness may have stemmed from both a sense of accomplishment in achieving a successful exchange and a feeling of connection with the robot. As exemplified by the relationship between the Sociable Trash Box and the children, there appears to be a state of well-being wherein one's abilities are utilized to achieve an active and joyful state.

BOX 4.6 PLAYFULNESS

Marius Hoggenmueller

Designing for playful interactions with robots challenges the prevailing utilitarian perspective on technology, embracing the potential to foster imagination, creativity, and emotional regulation. There is now a wide range of research prototypes and commercial robotic platforms that serve solely entertainment purposes, while others incorporate playful elements to achieve higher-level goals. Examples include social robots that provide companionship, for example, to mitigate loneliness among older adults, or those that are deployed in classrooms for educational purposes.

In the educational context, researchers are increasingly advocating for robot forms and interaction paradigms that do not replicate human tutors, for example, through a humanoid appearance or implementing conventional didactic methods (Zaga, 2019); instead, they argue for intelligent artifacts that allow for more **self-directed and open-ended learning** through play among children. These intelligent artifacts—sitting somewhere on the spectrum between toy and robot—often take on a low or non-anthropomorphic form.

Aesthetically and conceptually, they share strong similarities with the ludic design movement, characterized by open-endedness and ambiguity. The term "ludic design" was coined by Bill Gaver in the early 2000s and builds on cultural theorist Johan Huizinga's book "Homo Ludens" which explores the role of play in shaping culture and society (Gaver, 2002). Acknowledging humans as inherently playful creatures, Gaver demonstrated how designing domestic technologies following ludic design principles can **promote curiosity and exploration**, creating value that is neither purely utilitarian nor solely for entertainment. Like those found in interaction design more generally, ludic design principles in HRI (Lee & Jung, 2020) often include a defocus on external goals and making the machine, in this case the robot, not obviously perceptible as such. This broadens the scope for envisioning alternative robot appearances and the kinds of play that can arise from interactions between humans and robots.

WOODIE: A PLAYFUL ROBOT FOR PLACEMAKING IN THE CITY

Playgrounds, green spaces for recreational activities, public art installations, and festivals are just a few examples of how to encourage playfulness in the contemporary city. Yet, if we think about urban technologies, including automation and robotics, we mostly imagine services that increase efficiency, such

FIGURE 4.6 A child drawing a humanoid robot with chalk on the ground next to Woodie.

as through the roll-out of autonomous vehicles. In one of our research projects, we challenged this predominant belief about urban technologies and instead designed a playful, non-utilitarian robot that promotes creative placemaking in the city.

Following a designerly approach to HRI, we created Woodie, a slow-moving urban robot that draws on the ground using conventional chalk sticks

(Hoggenmueller & Hespanhol, 2020; Hoggenmueller, Hespanhol & Tomitsch, 2020). Capable of producing simple, pre-programmed line drawings such as flowers and love hearts, Woodie uses the public space as a large horizontal canvas. While urban environments are increasingly pervaded by digital screens and signage, our aim was to implement a low-tech approach to the dissemination of content and thereby make use of a robot's ability to physically manipulate the surrounding environment.

In terms of the robot's appearance and interactional qualities, we implemented ludic design principles. For example, in the case of Woodie's physical proportions, we followed the principle of de-familiarization by opting for slightly larger dimensions compared to common domestic service robots (e.g., vacuum cleaning robots). This would also take into consideration the scale of the city and increase the chance of the robot being noticed by passersby. To entice curiosity and allow for playful exploration, we further aimed to diminish any associations with what a robot should look like, behave, and be capable of doing. Thus, we opted for a non-anthropomorphic appearance and instead integrated a low-resolution lighting display in the robot's outer shell to attract and communicate with passersby through ambient visualizations.

The deliberate slowness in the robot's movement, the rendering of non-geometric line drawings that resemble the style of hand-drawn sketches, and the lack of any direct interaction channels all challenge the assumption that robots should be efficient, strive for perfection, and designed to serve humans for mundane tasks only.

Our exploratory design research project concluded with the deployment of Woodie at an annual light festival in Sydney, Australia. Over the course of three weeks, Woodie was drawing its sketches on a quiet laneway during the evening hours. Chalk sticks were handed out to surrounding people so they could extend and add to the robot's drawings. Not only the collaborative activity itself but also Woodie's presence and appearance, resulting in people perceiving it as a living being, attracted considerable attention and interest from the public. A mother who was visiting the site with her daughters commented on the potential of urban robots such as Woodie to encourage playfulness in cities: *"For my girls, it's fantastic because they could spend one hour being entertained by something like that. So, to me, that's fantastic."*

BOX 4.7 PARA-FUNCTIONALITY

James Pierce

In medicine, a **para-functional habit** is one in which a body part is used in an uncommon way. For example, any use of the human teeth or mouth to perform habits other than eating, drinking, or speaking is considered parafunctional. These include habits like fingernail biting or pencil chewing.

In design, Anthony Dunne and Fiona Raby have extended parafunctionality to advocate a critical, arts-inflected way of approaching the use of "functional" objects. Whereas traditional product design focuses on utility, the parafunctional design approach "attempts to go beyond conventional definitions of functionalism to include the poetic" and to "encourage reflection" from users about how technologies condition their behaviors (Dunne & Raby, 2001).

Put simply, parafunctional designs support unusual uses that **break with common notions of functionality**. In doing so, they can promote critical reflection and poetic, aesthetic experiences.

ROOMBA+CLIPS (AN ECCENTRIC SENSING DEVICE)

Both notions of parafunctionality—the medical and critical design usage—are evident in my project Roomba+Clips (Pierce, 2020, 2021). This eccentric device physically combines two different smart home products—the iRobot Roomba robotic vacuum cleaner and Google's Clips hands-free camera. These two products exemplify consumer applications of autonomous technologies, which employ digital sensors and AI to perceive their environments, make decisions, and perform actions. The Roomba operates autonomously in order to clean areas of the home that it thinks need cleaning. Clips operates autonomously in order to capture photographs it thinks you might like. Roomba+Clips mechanically couples these two products with a flexible plastic neck connecting the "body" of the Roomba with the "brain" of Clips (Scientists consider the eyes to be an external portion of the brain).

When activated, the Roomba bumbles about comically, while the Clips camera oscillates wildly—capturing short video clips—each time the vacuum strikes an object and changes direction. The newly formed product is vaguely anthropomorphic. In practice, it is highly engaging to nearby people, often

FIGURE 4.7 Roomba+Clips Accessory Kit.

inviting direct and playful interaction such as waving at the camera eye or placing a foot in front of the body so it changes direction.

But what is the use of Roomba+Clips, and its application of parafunctionality? At a basic level, this device employs Dunne and Raby's techniques of poetic expression and inviting critical reflection. Extending these uses further, I designed Roomba+Clips as an exploratory research tool for investigating design opportunities, limitations, and concerns connected to robotics and autonomous technologies. For example, conceptually, this project uses exaggerated forms to highlight unpredictable and emergent qualities of autonomous everyday technologies. Roomba+Clips is an embodiment of the myriad privacy violations and social tensions resisting with autonomous sensing technologies.

Empirically, my deployments of Roomba+Clips in public and exhibition spaces provide tentative insights into design techniques for mitigating, as well as exacerbating, these issues.

For example, audiences were typically drawn to Roomba+Clips, rather than threatened by it. They laughed, smiled, and even danced with it—all the while, assuming correctly that it was photographically monitoring and recording them. Roomba+Clips illustrates how autonomous technologies can "disarm with charm," and thus avoid transgressing social and personal boundaries. This technique might be used to improve the experiences for users and bystanders. Of course, it also holds great potential for abuse by masking or diverting attention away from privacy, safety, and exploitative concerns.

More subtly, the project illustrates how users and bystanders can both correctly and incorrectly infer the functionality of robotics through its form language. For example, many participants incorrectly inferred that the devices brain and body were digitally connected, when their dependence is in fact limited to mechanical connection. Conversely, participants correctly inferred that the lens opening and blinking lights indicated an active camera sensing system.

In these ways, Roomba+Clips demonstrates how a parafunctional object can be useful as a generative design and research tool, one that adapts and applies artistic techniques of provocation to both illustrate and develop hypotheses, insights, and design ideas.

BOX 4.8 RELATIONALITY

Dave Murray-Rust

Relationality is not a method as such, more a way of looking at the world. **Relational frameworks** have at heart the idea that the relations between things are as important as—if not more important than—the things themselves. While this may sound like an esoteric piece of thinking, there are many practical ways in which it plays out. As an example, the rise of "relational databases" that came to dominate information management was based on the power of making connections between elements of tables. Taking this approach a step further, the semantic web is founded on the idea that it is the relations between concepts that are important, and in many cases that concepts are defined as their relations.

What does this have to do with robots? One way of thinking about a robot is the set of functions that it can perform, its abilities, the tasks it can carry out, and the efficiency, accuracy, and speed with which it can perform them. There are many places where this viewpoint makes sense, particularly in a very production line-oriented setting, but it breaks down when we want to consider what happens as increasingly capable robots start to engage with humans in increasingly open situations. Encounters with robots are shaped by the way that people choose to respond to and interact with them as much as they are by the robots' actual capabilities. If a robot—or any technology—enters the workplace, a key determinant of its success is the way that it fits into and changes the web of existing practices, and how it meshes with the workers set of values about their work.

This is where relational frameworks come in. They direct us to consider the robot not as a discrete object, but to find it in the changes to **behaviors and practices around it**, just as humans are found not just in biology but also in the play of culture, society, needs, desires, and values that make up our lives.

LICHTSUCHENDE

We can illustrate some of these questions through looking at a concrete example—a "society" of robots called Lichtsuchende. These robots began in 2013 as an experiment into light-based interactions—homing in on sources of bright light using a set of simple photosensors—but quickly grew to have a crude psychology based very loosely on Maslow's hierarchy of needs. This resulted in a collection of static robots that looked around the space trying to

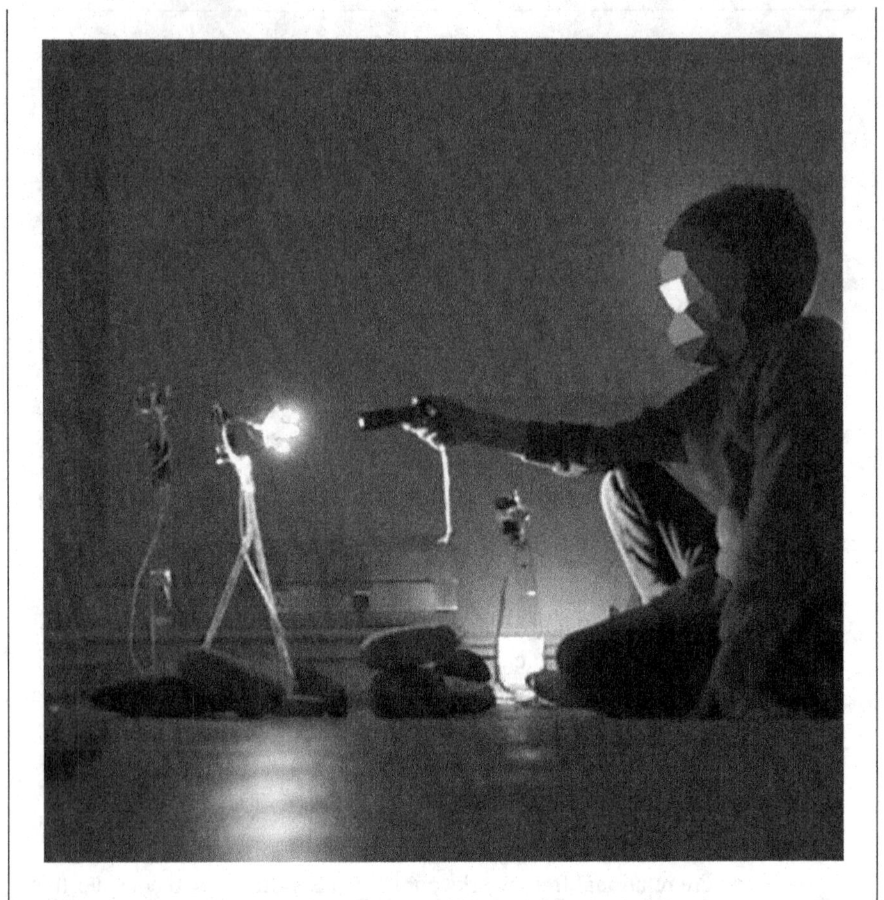

FIGURE 4.8 A small human interacting with the robots on their level.

find light sources to interact with. As with many collections of computational things, this led to the emergence of surprising, unprogrammed behaviors and constellations of action.

Relationality was central to the robots' activity—they did not have a function as such, and a single robot on its own would end up cycling through a set of movements, often becoming stuck in certain local maxima, such as staring at its own reflected light from a wall or ceiling. When robots were brought together, they had the possibility to enact their social relations, finding connections with each other, bonding, and sharing exchanges of light. Within each device, the psychology was a simple set of behaviors, with human-readable symbols to describe the state of the robot. When put into a space with other, similar robots, these psychologies could be enacted and played out, turning into a set of social relations, that saw patterns of movement shift around the collection of robots, groups form and dissolve, and subtleties of shared movement that were not purposefully built in.

We can also look at the relations between human visitors and the robots. Figure 5.8 shows a moment captured between two humans and a subset of the robotic environment. The small human is enacting their cultural behaviors—curiosity, exploration, and play. They are in the process of uncovering the ways that the robots respond, and trying to figure out what relation there is there: Are they collaborators, or responders? Pets or toys? This is happening in the context of the robots' unfolding relations, where the ongoing interaction between the group of three robots has been modulated by an excitable human, drawing much of the energy with rapid and mobile activity. The larger human clearly has a care relation toward the smaller one—their movement is protective, while leaving space, looking after development and safety at the same time.

These relations are played out physically in the proxemics between the participants—the small human entering the close zone around the robots, the larger one staying outside, but nearby.

The relations that play out go beyond immediate interactions. By confronting people with these playful robots that don't have a clear functional purpose, we were able to explore the attitudes and beliefs that come up around the technology. This shifts the default conceptions of a cyborg with a fixed grin or a Kuka arm with inhuman precision to open a space for negotiation of what robot-ness is. Visitors noticed different ways to describe the robots and their behavior from swarms to sea creatures, friendliness to thread displays. They explored different interactions from distant signaling to rolling on the floor between the robots. This opened a space to understand the robots in new terms, and to think about what relations to enact with them.

REFERENCES

Coeckelbergh, M. (2011). Humans, animals, and robots: A phenomenological approach to human-robot relations. *International Journal of Social Robotics*, *3*, 197–204.

Dennett, D. C. (1989). *The intentional stance*. MIT press.

Dörrenbächer, J., Löffler, D., & Hassenzahl, M. (2020, April). Becoming a robot-overcoming anthropomorphism with techno-mimesis. In *Proceedings of the 2020 CHI conference on human factors in computing systems* (pp. 1–12).

Dunne, A., & Raby, F. (2001). *Design noir: The secret life of electronic objects*. Springer Science & Business Media.

Frauenberger, C. (2019). Entanglement HCI the next wave? *ACM Transactions on Computer-Human Interaction (TOCHI)*, *27*(1), 1–27.

Frauenberger, C. (2021). What are you? Negotiating relationships with smart objects in intra-action. In M. C. Rozendaal, B. Marenko, & W. Odom (Eds.), *Designing smart objects in everyday life* (pp. 76–90). Bloomsbury Visual Arts.

Gaver, W. (2002). Designing for homo ludens. *I3 Magazine*, *12*(June), 2–6.

Giaccardi, E., & Redström, J. (2020). Technology and more-than-human design. *Design Issues*, *36*(4), 33–44.

Gollwitzer, P. M. (1999). Implementation intentions: Strong effects of simple plans. *American Psychologist*, *54*(7), 493.

Hassenzahl, M., Borchers, J., Boll, S., Pütten, A. R. V. D., & Wulf, V. (2020). Otherware: How to best interact with autonomous systems. *Interactions*, *28*(1), 54–57.

Hassenzahl, M., & Laschke, M. (2014). *Pleasurable troublemakers* (pp. 167–195). MIT Press.

Hoggenmueller, M., & Hespanhol, L. (2020, February). Woodie. An Urban Robot for Embodied Hybrid Placemaking. In *Proceedings of the Fourteenth International Conference on Tangible, Embedded, and Embodied Interaction* (pp. 617–624).

Hoggenmueller, M., Hespanhol, L., Tomitsch, M. (2020). Stop and Smell the Chalk Flowers: A Robotic Probe for Investigating Urban Interaction with Physicalised Displays. In *Proc. of the ACM CHI Conference on Human Factors in Computing Systems (CHI'20)*

Ihde, D. (1990). *Technology and the lifeworld: From garden to earth*. Indiana University Press.

Kaptelinin, V., & Nardi, B. A. (2006). *Acting with technology: Activity theory and interaction design*. MIT Press.

Laschke, M., Diefenbach, S., Schneider, T., & Hassenzahl, M. (2014, October). Keymoment: initiating behavior change through friendly friction. In *Proceedings of the 8th Nordic conference on human-computer interaction: Fun, fast, foundational* (pp. 853–858).

Laschke, M., Neuhaus, R., Dörrenbächer, J., Hassenzahl, M., Wulf, V., Rosenthal-Von Der Pütten, A., ... & Boll, S. (2020, October). Otherware needs otherness: Understanding and designing artificial counterparts. In *Proceedings of the 11th Nordic conference on human-computer interaction: Shaping experiences, shaping society* (pp. 1–4).

Lee, W. Y., & Jung, M. (2020, March). Ludic-hri: Designing playful experiences with robots. In *Companion of the 2020 ACM/IEEE international conference on human-robot interaction* (pp. 582–584).

Leontiev, A. N. (1975). *Activities, consciousness, personality*. Politizdat.

Nicenboim, I., Giaccardi, E., Søndergaard, M. L. J., Reddy, A. V., Strengers, Y., Pierce, J., & Redström, J. (2020, July). More-than-human design and AI: in conversation with agents. In *Companion publication of the 2020 ACM designing interactive systems conference* (pp. 397–400).

Nicenboim, I., Oogjes, D., Biggs, H., & Nam, S. (2023). Decentering through design: Bridging posthuman theory with more-than-human design practices. *Human–Computer Interaction*, 1–26.

Nicenboim, I., Venkat, S., Rustad, N. L., Vardanyan, D., Giaccardi, E., & Redström, J. (2023, April). Conversation Starters: How Can We Misunderstand AI Better? In *Extended abstracts of the 2023 CHI conference on human factors in computing systems* (pp. 1–4).

Okada, M. (2022). *Robot: For symbiotic relationship with humans* (in Japanese). University of Tokyo Press.

Pierce, J. (2020, July). Roomba+ clips cam: Exploring unpredictable autonomy in everyday smart systems. In *Companion publication of the 2020 ACM designing interactive systems conference* (pp. 317–320).

Pierce, J. (2021, June). Eccentric sensing devices: Using conceptual design notes to understand design opportunities, limitations, and concerns connected to digital sensing. In *Proceedings of the 13th conference on creativity and cognition* (pp. 1–14).

Reddy, A., Kocaballi, A. B., Nicenboim, I., Søndergaard, M. L. J., Lupetti, M. L., Key, C., … & Strengers, Y. (2021, May). Making everyday things talk: Speculative conversations into the future of voice interfaces at home. In *Extended abstracts of the 2021 CHI conference on human factors in computing systems* (pp. 1–16).

Rozendaal, M. (2016). Objects with intent: A new paradigm for interaction design. *Interactions*, *23*(3), 62–65.

Rozendaal, M. C., Boon, B., & Kaptelinin, V. (2019). Objects with intent: Designing everyday things as collaborative partners. *ACM Transactions on Computer-Human Interaction (TOCHI)*, *26*(4), 1–33.

Rozendaal, M. C., van Beek, E., Haselager, P., Abbink, D., & Jonker, C. M. (2020, November). Shift and blend: Understanding the hybrid character of computing artefacts on a tool-agent spectrum. In *Proceedings of the 8th international conference on human-agent interaction* (pp. 171–178).

van Beek, E., Giaccardi, E., Boess, S., & Bozzon, A. (2023). The everyday enactment of interfaces: a study of crises and conflicts in the more-than-human home. *Human–Computer Interaction*, 1–28.

Verbeek, P. P. (2015). Beyond interaction: A short introduction to mediation theory. *Interactions*, *22*(3), 26–31.

Wakkary, R. (2020). Nomadic practices: A posthuman theory for knowing design. *International Journal of Design*, *14*(3), 117.

Willerslev, R. (2007). *Soul hunters*. University of California Press.

World Health Organization. (2010). Global recommendations on physical activity for health (p. 60). http://www.who.int/dietphysicalactivity/publications/9789241599979/en/

Yamaji, Y., Miyake, T., Yoshiike, Y., De Silva, P.R., & Okada, M. (2011). STB: Child-dependent sociable trash box. *International Journal of Social Robotics*, *3*(4), 359–370.

Zaga, C. (2019). Designing the future of education: From tutor robots to intelligent playthings. *Tijdschrift voor Human Factors*, *44*(3), 5.

5 Designing Robotic Imaginaries

Narratives and Futures

Cristina Zaga
University of Twente, Enschede, Netherlands

Design is a political future-shaping activity (Fry, 1999). We design technology that was not here before because we wish to change or improve the current state of things. Designers, engineers, and roboticists imagine, ideate, and develop technology—implicitly and explicitly—from a specific narrative (i.e., shared stories we use as a society to make sense, learn, and act in the world [Fisher, 1985]), set of personal and societal values, worldviews, and power structures (Mazé, 2019). Narratives, personal and societal values, worldviews, power structures, and norms contribute to developing specific imaginaries, i.e., *"collectively held, institutionally stabilized, and publicly performed visions of desirable futures, animated by shared understandings of forms of social life and social order attainable through and supportive of, advances in science and technology"* (Jasanoff & Kim, 2019, p. 4).

What are robot imaginaries, then? How do they shape our futures?

The idea of a mechanical agent preceded the actual implementations of robotic technology and the field of HRI (Gasparetto, 2016). Although "robot" is a 20th century concept, various robotic imaginaries have developed throughout our societies and contributed to shaping robot realities. While Western historians situate early ideas of the mechanical devices or machines self-operating (often called automaton/automata) back to the medieval craftsmen who engineered self-propelled machines (Stephens & Heffernan, 2016), the idea of an artificial entity moving autonomously can be traced back to early civilizations and has a tight connection with the desire of humans to connect to the divine and surpass humans' perceived and factual limitations. Greek, Islamic, and Buddhist civilizations envisioned automated creatures as projections of humans into the religious realm, creatures that bring salvation and extraordinary power to humans' lives (Trovato et al., 2021), providing early metaphors of a higher intelligence.

Later in (Western) history, automata actively explored the relationship between being human and being artificial, between agency and movement, and between human limitations and artificial enhancement of rationality (Stephens & Heffernan, 2016). Automata were displayed in fairs and at courts and attracted attention for their

DOI: 10.1201/9781003371021-5
This chapter has been made available under a CC-BY-NC-ND license.

mechanical sophistication, "illusion of life," and intelligence. Therefore, automata carried specific perspectives and societal meaning; they were "arguments as well as amusements" (Schaffer, 1994, p. 16). Particularly in the 1800s, automata represented arguments from mechanistic philosophy (i.e., a philosophy that reduces the universe to mechanical principles) almost to reduce humans to machines and—conversely—anthropomorphize machines. Most of the automata of those times were carefully crafted machines celebrated as extraordinary technological "curiosities" (Stephens & Heffernan, 2016). In some cases, automata pushed the illusion of life and intelligence to explicit deceptions, like in the case of the Mechanical Turk. This chess player machine was remotely controlled by humans and hidden in the machine.

With the advent of the Industrial Revolution in the West, automated machines started to embody the utilitarian desire for efficiency, effectiveness, and productivity for economic gain, redefining what it means to be human (which can be as well considered an imaginary in itself) while dehumanizing human activities and introducing forms of artificial slavery (Hampton, 2015; Rhee, 2018). Notoriously, the term robot is intrinsically connected to work labor as it is a Czech translation of forced labor of bondman, and this definition has impacted the developments of robotic devices as utilitarian and servile machines both for industry and the military. Automation has thus transformed cultural imagination to the extent that scholars identified the rise of a "mechanical age" (Carlyle, 2015) without slavery. Understanding the boundaries between humans and tools becomes thus fundamental, as well as examining the (capitalistic) drivers of the mechanical age: robot imaginaries connect more and more with societal structures and structures of power and oppressing linking with issues of race, slavery (and particularly in the U.S. with slaves' identities), and workers' liberation (Hampton, 2015).

An example is Lang's movie Metropolis, which popularized critical reflections on industrialization (and the idea of robots taking over work), painting the "maschinenmensch" (i.e., a robot capable of work) Maria as a powerful and sinister force bringing chaos to society, ammunition to the mechanization of society, and the blind fascination with societal progress intended as the technological advance of capitalism (Babich, 2012; Huyssen, 1986; Minden & Bachmann, 2002).

In the 20th and 21st centuries, we can see how past imaginaries, values, worldviews, politics, and cultures have shaped the hegemonic robot imaginaries and technological developments. Robots are staples in mass media, science fiction, and technoscientific scenarios that actively shape how we design robots. Images of robots alternate from perilous all-mighty machines willing to take over our jobs and lives (e.g., Terminator) to benevolent butler companions or sidekicks helping us (e.g., C3PO, R2D2) to super-intelligent "god-like" entities benefiting humans in the name of progress. Scholars have shown that current robot imaginaries (particularly in Western, educated, industrialized, rich, democratic countries, aka WEIRD) translate into designs that often fixate on utilitarian functions and misrepresent the technology's capabilities, focusing on fully autonomous life-like agents (Alves-Oliveira et al., 2021). Moreover, robots reinforce stereotypes, augmenting biases and becoming renewed forms of marginalization (Zaga & Lupetti, 2022). They tend to feature predominantly humanoid forms, gender-normative, and white-Caucasian appearance and behavior (Cave & Dihal, 2020), and many scholars and designers are finding ways to elaborate alternative imaginaries (Auger, 2012, 2014; Cheon & Su, 2018; Lupetti et al., 2023; Luria & Candy, 2022).

Many of the current status quo, beliefs, and wishes about robotic technologies are thus shaped by centuries of stories that people have told about humans and machines and our relationships. Therefore, the stories of researchers, designers, policymakers, and the public speak about how robots matter and influence visions of what we consider the future. How we frame and envision the future has crucial social, ethical, and political consequences. How we imagine future robotic technology thus influences how we act in society to what is possible and desirable (Pelzer & Versteeg, 2019). Therefore, human–robot interaction (HRI) designers hold considerable societal power and responsibility, sometimes unbeknownst to them (Verbeek, 2006).

We, as HRI designers, envision and design robots implicitly and explicitly, imbuing values, culture, norms, narratives, and power structures in their appearance, behavior, role, and functions (Cheon et al., 2021; Luria et al., 2020; Luria & Candy, 2022; Zaga & Lupetti, 2022). How a robot should look, behave, and its role in an interactional context carries specific imaginaries and contributes to the mutual shaping between technology and society (Šabanović, 2010). Therefore, the future-shaping orientation of design has an ideological and political component that needs to be critically examined in the design process by researchers, designers, and stakeholders.

At the intersection of the design and HRI, a growing body of work explicitly addresses how we shape robotic imaginaries narratives to develop alternative imaginaries and futures (Auger, 2012, 2014; Cheon & Su, 2018; Lupetti et al., 2023; Luria & Candy, 2022) through design futuring. Building upon Kozubaev et al. (2020), we define design futuring *as "a variety of approaches that leverage design to explore futures as a means to comment on—and potentially change—the present"* (p. 2).

Diverging from the dominant positivistic tradition of HRI, design futuring approaches focus on critique, provocation, and challenging the status quo producing intermediate-level knowledge (Lupetti et al., 2021) and connecting to meaning-making modes of knowledge production in the humanities and social sciences. As such, connecting with approaches such as critical design, adversarial Design, speculative design all geared to examine and challenge assumptions, values, ideologies, and socio-behavioral norms embedded implicitly and explicitly in design (Bardzell & Bardzell, 2013, DiSalvo, 2012, Dunne & Raby, 2024).

While these approaches have specific foci and methods, a common denominator is going beyond solving problems by generating artifacts, instead asking questions, provoking reflection, and challenging the status quo (Kozubaev et al., 2020) by generating artifacts, narratives, or by stimulating tangible reflections. As Dunne and Raby (2024) suggested, design open windows on the future to understand the present better. Through engaging in material reflection or artifact generation, it can *"bridge the experiential gulf between inherently abstract notions of possible futures and life as it is apprehended, felt, embedded, and embodied in the present and on the ground"*

(Candy & Dunagan, 2017, p. 137) and in so *"Imagining a future in enough closed-down specificity that we can grasp and experience aspects of it in the present moment, while also opening up to divergent experiences and reactions of the design artifact in use"* (Kozubaev et al., 2020, p. 4).

Design futuring approaches challenge the methodological status quo of HRI as well. It invites researchers, designers, and practitioners to engage in reflexive approaches, which are not yet established in the field. By shedding lights on the responsibility in design as well as the socio-economical drivers behind the developments of robots, design futuring approaches help HRI gain a meaningful understanding of the ways in which technology could be designed otherwise in line with ethics, values, and social justice concerns. HRI practices add a layer of complexity to design futuring as well, by engaging formally with agentic technology that has a more apparent relational facet than other artifacts.

At the same time, there is not a univocal "future" that brings "progress," there is not one timeline from the "now" to the "future" but a multitude of narratives, perspectives, timelines, and points of view that should be examined and materialized, critically (Escobar, 2018). Therefore, both HRI and design futuring should engage in pluralizing practices, including in their processes the people for whom the impact of the HRI design is the greatest, connecting with value-oriented participatory approaches (such as pluriversal design) taking power-dynamics into account to design for socio-cultural change.

In the following sections, we provide a set of approaches, methods, and practices to challenge current robot imaginaries by examining, analyzing, and generating narratives and futures.

In Box 5.1, Cheon and Su elaborates on Futuristic Autobiographies, FAB, a method to help roboticists (and potentially HRI stakeholders) to be aware of their values and examine how values intermingle with design practice. Using a diegetic device—futuristic autobiographies—researchers are encouraged to analyze past practices with future lenses. Using the format of autobiographies helped roboticists to examine disciplinary assumptions as well as tackle visions of the future through a fictional device.

In Box 5.2, Gamboa and La Delfa propose Soma Design as a first-person perspective method to turn inward and center on the individual experience of the designer to make sense of values, worldviews, and meaning. The authors used Soma Design to analyze the lived experience and meaning behind human–drone interactions. At large, Soma Design could support deeper reflexive and aesthetic engagement with a plurality of experiences to make sense of and materialize HRI futures.

In Box 5.3, Alves-Oliveira champions the concept of Metaphors as a productive tool to engage in the co-sense making of values, narratives, and meaning in HRI. Metaphors can generate alternatives to current narratives and the status quo with designers, roboticists, and stakeholders. The section hones in a card-based tool for team explorations of alternative metaphors for HRI.

In Box 5.4, Rebaudengo details what Design Fiction is and how it can be used to examine technology critically. Design Fiction makes use of diegetic (i.e., narrative-oriented) prototypes to create detailed scenarios and tangible objects to envision, discuss, and understand how futures might unfold, particularly in the context of emerging technologies and societal changes. To illustrate how design fiction can be implemented, Rabaudengo details a design fiction prototype, the Teacher of Algorithms, a disciplinary character in a world where technology needs behavioral training.

In Box 5.5, Auger centers on the Speculative Design approach, detailing two strategies: speculative futures and alternative presents. Speculative design materializes desirable futures and broadens our horizons by generating artifacts. It encourages us to think of alternative futures and discuss our direction. The section features a speculative design project developed before the widespread development of domestic smart speakers. Real Prediction Machines explores the domestication of robots and the impact of data collecting/parsing agents in our everyday lives, stimulating reflections and provoking us to think of alternative imaginaries.

In Box 5.6, Lupetti delves into Adversarial Design an approach to design futuring focused on unveiling and engaging with the politics of design while actively challenging the status quo. Design becomes a vehicle to make hegemonic structures that influence decisions and behaviors apparent. To illustrate Adversarial Design, Lupetti introduces Steering Stories to challenge the current narratives of robot cars through contestational artifacts that make narratives and frictions of autonomous driving apparent.

In Box 5.7, Luria details Ethnographic Experiential Futures (EXF), a method of participatory foresight from the Future Studies tradition to identify current visions of the future within a particular community or group and using these visions as a foundation for creating tangible and rich probes of the future, which can be used for reflection. EXF has been used in the project Letters from the Future to elicit and tackle ethical tensions in the design process.

In Box 5.8, Mi proposes Blending Traditions to underline the importance of cultural traditions and imaginaries in the HRI design process, a sort of cultural etiquette that needs to be considered. To illustrate the concept, Mi introduces the Moja Robotic Chinese Orchestra, an artifact that embodies cultural blending. It incorporates traditional Chinese cultural narrative methods into performance design. Moja robots serve as both clues in the narrative and promoters of scene transitions and storyline progression, collaborating with human actors to present sophisticated performances on a limited stage, thereby providing reflection-in-action about the cultural shaping of HRI.

BOX 5.1 FUTURISTIC AUTOBIOGRAPHIES

EunJeong Cheon and Norman Makoto Su

Futuristic autobiographies (FABs) are an empirical method to augment semi-structured interview studies. Situated in deep fieldwork and other ethnographic methods, they are fruitful ways to understand the values that are ingrained in groups. FABs are a **value elicitation tool** for producing qualitative data which can then be analyzed appropriately (e.g., grounded theory analysis). By analyzing FABs, one can answer research questions about those involved with technology (e.g., designers, developers) that are difficult to ask directly, such as: "What values do you hold?" and "How do these values intermingle with your practices?"

FABs involve the presentation of stories, or prompts, created by the researcher which are grounded in background research (e.g., analysis of fieldwork or archival data) that involve the participant as a key character (hence, the autobiographical nature of the exercise). These stories take place in the future, and participants are asked to think about the "past" of futuristic stories; in other words, they are to complete their own, albeit fictional, "autobiography." Rather than ask about current practices or viewpoints, FABs challenge informants to weave narratives *featuring themselves* as a main character. Inevitably, this act of storytelling leads informants to incorporate aspects of their lives—their practices, challenges, experiences, philosophies, etc. Although not easy, from these autobiographies, we can gain insight into the values these informants hold about the technologies they use, design, or theorize. Most significantly, akin to diegetic prototypes in design fiction, FABs create a user who is diegetic. This **diegetic user** is situated in the futuristic autobiographies (often along diegetic prototypes). By putting forward to participants fictional future situations, our method opens a space for discussion (Cheon et al., 2019; Cheon & Su, 2017, 2018).

CONFIGURING THE USER: "ROBOTS HAVE NEEDS TOO"

We employed FABs with 23 HRI researchers (hereafter called roboticists) whose research involved humanoids. FABs were incorporated into our semi-structured interview protocol and were conducted with individual participants in-person as part of the interview. FABs were introduced to participants with the following prompt: *"I'm going to present you with a set of stories about your future with robotics, each of which will be followed by some questions. Feel free to use your imagination in your answers. There's no right or wrong answer."*

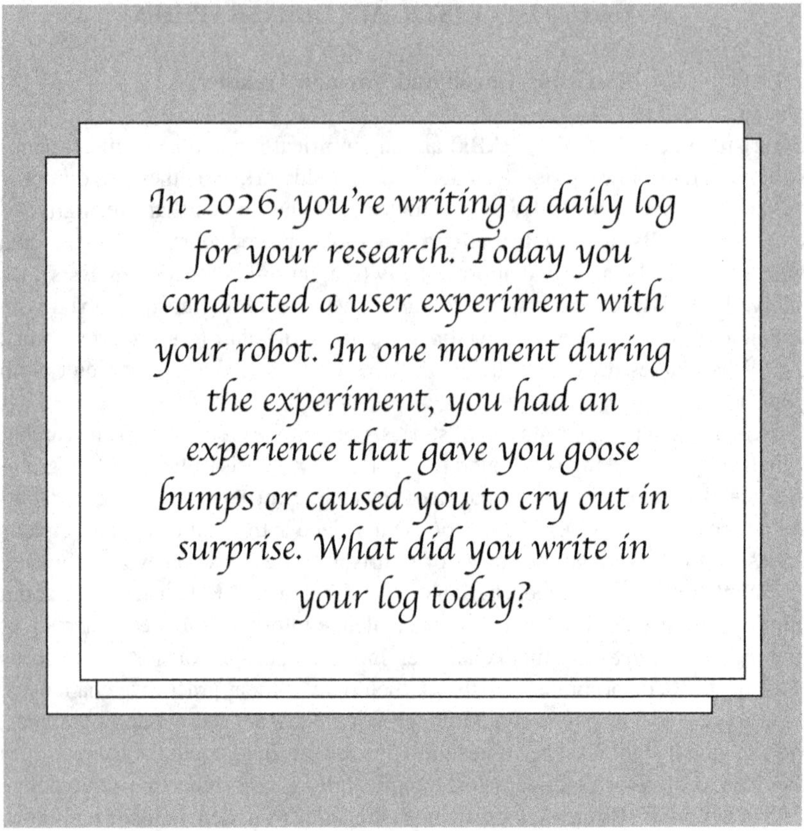

In 2026, you're writing a daily log for your research. Today you conducted a user experiment with your robot. In one moment during the experiment, you had an experience that gave you goose bumps or caused you to cry out in surprise. What did you write in your log today?

FIGURE 5.1 An example of futuristic autobiographies (FABs) prompts used in our study.

The prompts were then given to participants one-by-one on a piece of paper. The participant was guided to respond and discuss their FABs orally. For example, one FAB prompt (Figure 5.1) sought to have participants articulate what constitutes surprise in terms of a technological discipline (i.e., robotics). By revealing what is unexpected or shocking, this FAB could conversely uncover what constituted the expected, mundane, or routine in robotics. In this sense, the FAB served as a circuitous route to get what we wanted.

We asked follow-up questions on aspects (e.g., probing on details and motivations) of their FABs. Sessions lasted 30–70 minutes and were conducted at multiple sites (a major HRI conference and two university robotic labs). Our initial research question (i.e., what perspectives and values might roboticists have when developing future robots?) evolved throughout the collection and analysis of data. When we reached data saturation in our coding process, our findings illustrated how differently roboticists imagine the current and future robot user (Cheon & Su, 2017). Given the diverse background of researchers doing HRI, the FABs reflected a multitude of different values from one individual to another. Grounded theory allowed us to focus on finding common themes with respect to values across our FABs.

Autobiographies allowed roboticists to discuss values in their own terms. Roboticists were able to—implicitly and explicitly—describe assumptions of their disciplinary practices with technology, values held by non-technologists and other users and their interactions with such values, societal challenges to their work, and future, ideal directions for their technologies in the world.

These autobiographies were speculative but, in the tradition of design fiction, entirely plausible and grounded in their own experiences.

Although FABs proved invaluable to our studies on roboticists, it remains to be seen if futuristic autobiographies can be beneficially applied to other domains. However, we believe this method has great promise to augment interview studies that align with Value-Sensitive Design's empirical approach to discovering values. When FABs are carefully crafted from background research, they create rich interviews where informants speak for themselves on how their practices and values are intertwined now and in the future.

Another project of ours seeks to explore how FABs can be used in a collaborative setting among various stakeholders (Cheon et al., 2019). We were particularly interested in whether the sharing of FABs would allow participants to better collaborate with each other and reflexively consider and incorporate differing values of all participants. We hope researchers utilize FABs and account for their strengths and weaknesses so as to suggest further refinements.

BOX 5.2 SOMA DESIGN

Mafalda Gamboa and Joseph La Delfa

Soma Design aims to create *"applications where the interactions subtly supports users' attention inwards, towardtowards their own body, enriching their sensitivity to, enjoyment and appreciation"* (Höök, 2018). Basically, if you can develop your ability to reflect on what you are "feeling" you are better able to appreciate and create your lived experience. Both as a designer and a user. Why is this useful? Well, every individual has a unique interpretation of the world, amassed from a unique permutation of lived experiences. The search for a universal truth often weeds out these interpretations that can lead to novel designs (Winkle et al., 2023). While User-Centered Design and Participatory Design are important approaches that should not be neglected, we emphasize that a **first-person perspective** must feature more prominently in HRI for it to truly embrace/realize its interdisciplinary status. Soma design has the potential to enable HRI community to tap into a wealth of novel, contextually suited, and culturally appropriate design outcomes.

This means embracing first-person design and analysis methods (Höök, 2018). Get started by moving and living with your chosen technology and paying attention to what you feel and how you feel it—keep records of these feelings (like videos, photographs, journals). Common ways of keeping these records including plotting the "trajectory" of your experience over time, or drawing your experience on a Body Map can help tease out what happened (Tennent et al., 2021). While discussing your experiences with a somatic connoisseur, people with in-depth knowledge of a sport or bodily practice (like yoga or dance) can give you a different perspective on how it happened (Höök, 2018).

Soma Design is not a method that must be followed "by the book" in order to achieve a desired outcome. Accessing **felt experiences** and using them to inform your design happens differently for everyone. Therefore, the methods and concepts referred to in this section are not to be understood verbatim, but experienced in practice, tried out. In Soma Design, you should bend the rules and enjoy doing so.

SOMATIC DRONES

Soma Design has inspired our work (Gamboa et al., 2023; La Delfa et al., 2020)—we have created experiences with them where the body takes center stage. For example, Drone Chi (Figure 5.2) is a human–drone interaction experience that is designed to bring attention to the way you move, breathe, and balance yourself (La Delfa et al., 2020).

FIGURE 5.2 Drone Chi, human-drone interaction experience.

Through the simple act of moving a drone with your hands, smooth and meditative movements are facilitated. The experience unfolds in four stages. First, you pick a drone from a vine. Once it's in the air, it hovers in place so you can explore the relationship between your hands and the brightness of the drone. Once you are comfortable with this relationship, the drone then begins to fly in a slow, vertically oriented circle. If you keep the light bright, the circles grow larger and the drone gradually turns pink. Reactive to your movements, you can now move it around the room and explore the relationship between your body and the drones.

Most importantly, our own engagement with the drones as a design material was weaved into our process. First-person narratives, as we offer below, are part of understanding and describing somatic interactions:

"The first time I flew a drone that responded directly to my hands, one aspect quickly stood out to me. A feeling of being intimately coupled to the drone. As if every part was playing a crucial role in keeping the drone in the air. I felt arrested, but not under control. The sensitivity of the drone to my body was matched by my body's sensitivity to the drone. It was like carrying a cup of hot tea with a book under your arm, any sudden movement from any part of your body could be seen in the tea cup and vice versa. This was an experience that I wanted to understand more deeply, to realize the potential design outcomes that could come from it. So, alongside daily flights with the drone, I took Tai chi classes.

Together, these practices allowed me to explore lots of potential designs. In these classes, I found that the feeling of smooth, coordinated movement and the imagery that was invoked in the class (e.g.: movement descriptions such as holding the ball of energy) were important experiences to draw form when moving with the drones.

Finally, using an industrial design approach, I narrowed the open-ended explorations into Drone Chi. In doing so, I leveraged my first-person experiences of flying drones and learning Tai Chi to create a unique design outcome" (La Delfa et al., 2020).

BOX 5.3 METAPHORS

Patricia Alves-Oliveira

Metaphors create comparisons between two different objects, people, or things. They describe something by saying it is something else. An example of a metaphor occurs when saying "you are my sun." Metaphors are a particularly relevant method to use at the start of projects when the design space is still open. They support the generation of innovative ideas for problem definition. Generating multiple ideas from the beginning opens opportunities to discuss multiple, possible alternatives for a problem. This supports the commitment to a vision for the project after having discussed pros and cons of multiple other ideas.

Metaphors are an ideal method for multidisciplinary teams to work together. They help **create shared meanings** for complex and abstract concepts, such as time and emotions, and include discussing trade-offs from the perspective of each team.

In the field of robotics, metaphors are a means that can support teams of designers, engineers, social scientists, and others to work together toward the creation of a robot's role, behavior, or embodiment. It can also be used to involve non-roboticists, such as the general public, to contribute their thoughts on how a robot should be created.

It is common to seek novel or alternative metaphors. This is especially true when needing to innovate and improve limitations of current robots. For example, while in the past the prevalent metaphor was "robots as servants," there has been a shift toward "robots as sidekicks." Both are transformational metaphors that lead to the creation of robots with different roles and behaviors when interacting with human counterparts.

METAPHOR FOR HUMAN–ROBOT INTERACTION

The "Collection of Metaphors for Human–Robot Interaction" (Alves-Oliveira et al., 2021) is a conceptual project aimed at identifying robot stereotypes and using metaphors to create novel ideas for how robots should behave, what role they can have in society, and how their embodiment should look like. This project was a multidisciplinary effort that involved a total of 28 experts from different fields, including engineers, designers, artists, filmmakers, philosophers, computer scientists, and psychologists, among others, to think together about this problem.

FIGURE 5.3 A possible human-robot interaction inspired by a tumor as a relational metaphor.

The project made use of the New Metaphors Toolkit (Lockton et al., 2019): a set of 150 cards, including 75 "Thing 1" (type of cards with images) cards and 75 "Thing 2" (type of cards with text) that can be combined to produce thousands of possible combinations. This method can either be used in-person with physical cards, or online with a card generator (https://www.michalluria.com/metaphors/).

A three-stage design exploration was used. In Activity 1, participants were invited to individually generate as many metaphors for HRI as they could. We aimed at eliciting wild ideas for the metaphors, valuing a fast-paced activity with an emphasis on quantity over quality to kick off the project. The value of this individual activity was to avoid external influences and bring each own unique and raw ideas. The result was the generation of a total of 21 different metaphors for robots.

In Activity 2, participants gave the metaphors they generated individually to another participant, and the goal was to have participants generating metaphors for concepts created by others. This enabled building on each other's ideas. For example, while a participant in Activity 1 generated the idea "How can robots as shadows be a metaphor for embracing the duality of human–robot co-existence?" in Activity 2, another participant generated the metaphor "How can robots as shadows be a metaphor for a robot that follows you around anywhere?" A total of 31 new metaphors were generated during this stage.

Activity 3 had the goal of diving deeper into one metaphor to create a detailed interaction between a human and a robot where this metaphor was explored. This was a group activity, and each group decided on a metaphor to explore. Groups started by unpacking their idea by discussing the characteristics of the metaphor they were working on (see Figure 5.3 for one example), and by reflecting on the concept of robots they wanted to promote. After this, they created a storyboard detailing how this metaphor for HRI would be applied to a real-world scenario.

As emerged from this project investigation, exploring unconventional robot metaphors, such as robots that make mistakes or act unpredictably, leads to profound considerations of the social, cultural, and ethical implications of robotic artifacts. In particular, the New Metaphors Toolkit provides a framework for roboticists to break free from preconceived notions and broaden the design space of robots. This project is meant to inspire roboticists, designers, artists, linguists, and engineers to develop robots that far surpass our current assumptions by acquiring different roles, shapes and forms, and interaction modalities with humans.

BOX 5.4 DESIGN FICTION

Simone Rebaudengo

Design Fiction is a creative and speculative approach that blends design and storytelling to explore and prototype possible futures. By creating detailed scenarios and tangible objects, this methodology offers a way to envision, discuss, and understand potential futures, particularly in the context of emerging technologies and societal changes. Design Fiction is not about predicting the future, but about exploring a range of possibilities, exploring implications in the everyday and often mundane and forgotten details of our lives, and often challenging our assumptions and expectations about the future.

At its core, Design Fiction involves creating physical prototypes, digital simulations, or narrative artifacts that are "real enough" to suspend disbelief. These artifacts, whether they are products, services, or environments, are set in a future world, complete with detailed contexts and stories. This immersive approach allows designers, stakeholders, and the public to engage more deeply with the speculative scenarios, bringing to life potential outcomes, opportunities, and risks associated with new technologies or societal trends.

One of the key strengths of Design Fiction is its ability to make abstract or distant future scenarios more tangible and relatable. By grounding speculative ideas in concrete objects and narratives, it bridges the gap between theoretical discussion and practical understanding. This is particularly valuable in fields like technology, product innovation, urban planning, and policy-making, where decisions made today can have far-reaching consequences.

Design Fiction prototypes can range from simple sketches or mock-ups to fully functional prototypes, depending on the purpose and audience. They are often **accompanied by narratives or story-worlds** that provide context and meaning. This narrative aspect is important as it frames the artifact within a plausible future scenario, encouraging viewers to consider not just the object itself, but its implications and the societal dynamics that might lead to its creation.

Overall, Design Fiction is a powerful tool for exploring **"what if" scenarios**, stimulating discussion, and fostering a forward-thinking mindset. Making the future more tangible enables designers, companies, and policymakers to anticipate and shape the future in more informed and responsible ways.

THE TEACHER OF ALGORITHMS

The Teacher of Algorithms (Figure 5.4) explores a not-so-distant or almost a parallel present shaped by the proliferation of intelligent devices and exploring how human roles might evolve in response to the increasingly autonomous technology in our lives.

FIGURE 5.4 A video snapshot from the teacher of algorithms.

The project was commissioned by ThingTank, a collaborative effort among researchers in design, anthropology, and computer science from various universities across Europe, Japan, and the U.S. in 2015, with the aim of offering a perspective in the form of design fiction prototypes and scenarios to critically reflect on the trajectory of our increasingly automated world.

"The teacher" is a fictional character, living in the back alleys of a megacity. He is engaged in the curious job of educating smart products such as robotic vacuum cleaners, learning thermostats, and other AI-driven devices. These devices, while advanced, still require a human touch to refine and improve their behaviors and functionalities. The teacher's role is to train these machines, not just in a mechanical sense, but in a way that imbues them with a more nuanced understanding of human needs and environments.

In this world, smart devices are not just tools; they become entities with a need for behavioral training. When a device's machine learning algorithm fails to align with its owner's expectations or needs, it is sent to the Teacher of Algorithms, something similar to a trainer for misbehaving pets. This teacher employs various techniques, from digital punishments and rewards to sophisticated simulations of the device's home environment, to recalibrate and retrain these devices.

To create this short story and scenario several artifacts had to be created to fill in the teacher's world: from training dust and sticks to train robotic vacuum cleaners, to Pavlovian interfaces to train fans and thermostats.

The project was shown in several design and film festivals, it won an award at the Robot Film Festival and was part of a global traveling exhibition "Hello Robot" curated by Vitra Design Museum. "The Teacher of Algorithms" is not just a fictional narrative but a commentary on current and future challenges in the world of smart technology.

BOX 5.5 SPECULATIVE DESIGN

James Auger

Design typically takes place within relatively stable cultural and economic value systems with societal change happening through iterative advances in technology. This situation creates a certain set of limiting conditions—or constraints, that act to narrow the paths of possibility and support certain problematic forms of (design) practice. Speculative design emerged in the mid-2000s as a way of challenging the roles, methods, and purposes of (commercial, global Northern) design.

The approach has two key strategies: Speculative futures take advantage of the (relatively) predictable and iterative nature of technological development to speculate on near-future products and systems. In this case, the designer typically collaborates with scientists researching a specific emerging technology (for example, artificial intelligence, robotics, synthetic biology, etc.) then extrapolates its potential to inform the speculation. The designer effectively bridges the void between the disparate habitats of the laboratory and everyday life, exposing scientific research to the complex needs and desires of people. Alternative presents are speculative proposals that question existing paradigms through the use of different ideologies to those currently directing product development. These are speculations on **how things could be had different choices been made in the past**. The use of alternative value systems, a non-additive technological function and the removal of the constraints imposed by history to allow different approaches to emerge that were nullified by the dominant, hegemonic or "standard" narrative(s). Alternative presents can open up valuable future paths and create space for rich new imaginaries to flourish.

Much design practice remains driven by the motivations of the 20th century and as such products and systems are commonly designed and evaluated by outdated or inappropriate means of measurement. By isolating the designer from the constraints typically imposed by the design industry's relation to the market, it becomes possible to develop new forms of practice, more aligned with the complex issues we face in the 21st century.

REAL PREDICTION MACHINES

The Real Prediction Machines (RPMs) were developed after the completion of a PhD that examined potential routes for the domestication of robots. The premise was that the fictional and mythical image of robots continues to influence their research and development, in turn leading to forms and functions that are essentially maladapted for the domestic habitat—if robots are to enter our

FIGURE 5.5 Real prediction machine.

homes, we must begin by dismantling the stereotypical and romantic concepts that pervade. Related technologies can then be reconsidered via the theory of domestication (Silverstone & Hirsch, 1992). The project was developed in collaboration with Subramanian Ramamoorthy, a Professor of Robot Learning and Autonomy at the School of Informatics, University of Edinburgh.

The RPM project was based on technologies of sensing, machine learning, data mining, and algorithmic prediction. Normative robotic forms (anthropomorphic, zoomorphic) and stereotypical functions (butler, companion) were avoided to better align with the aesthetic and operational expectations of domestic products, while playing to the strengths of the technology. The aim was to communicate the experiential potential of Big Data and predictive algorithms by shifting their function from professional contexts, such as weather forecasting, structural engineering, or banking, to the context of everyday domestic life.

The project was developed just before the launch of Amazon's Echo—and likewise, it listens into domestic conversations, gathers data, and (machine) learns from the patterns that can ultimately be detected. These can then be used to make increasingly accurate predictions on specific events, such as when a couple might have their next domestic argument or the likelihood of the onset of a pre-defined chronic illness.

Once the predictive event has been chosen, the necessary data streams are identified, including, for example, outputs from live sensors, RSS feeds, and historical information. These continuously feed the prediction algorithm—the output of which controls a visual display on the prediction machine informing the viewer if the chosen event is approaching, receding, or impending.

Such technologies arrive in our everyday lives almost surreptitiously, for example, through small iterations in existing product lines. New untested services and functions become available, transforming various aspects of our lives—and only when they become mainstream do we begin to analyze their impact. Regarding big data, Mayer-Schönberger and Cukier (2013) pose some poignant questions: *"As big data transforms our lives – optimizing, improving, making more efficient, and capturing benefits – what role is left for intuition, faith, uncertainty, and originality?"* The purpose of RPMs is to begin questioning these technologies before they become an everyday reality.

BOX 5.6 ADVERSARIAL DESIGN

Maria Luce Lupetti

Coined by Carl DiSalvo (2012), the term Adversarial Design describes a perspective and related approach that looks at the **political role of design** and sets out to challenge traditional norms and the status quo of design by making use of provocative artifacts.

The approach breaks disciplinary boundaries and nurtures a kind of cultural production that sits at the intersection of design, art, engineering, computer science, and philosophy. Through the semantics of designed things, this approach engages with collective issues and creates spaces for agonism—a condition of disagreement and contestation. Adversarial Design, then, leverages the design medium to express dissent and to **create moments of contestation**.

In doing so, the approach manifests and emphasizes the political role of design and its entanglement with matters of public concern. It sets out to contribute to nurturing a democratic culture, in which disagreement and confrontation are constructive elements of public participation.

In Adversarial Design practices, information design can become a medium to reveal hegemonic structures, product design can emphasize neglected qualities and features to unveil normative configurations of things, and ubiquitous technologies can be a medium to amplify expressions of dissent in public arenas. DiSalvo (2012) describes these as **contestational artifacts**—designed objects that manifest aspects of a political condition and offer alternatives to dominant practices and agendas for the public to engage with.

Adversarial Design practices, however, extend far beyond the material craft of contestational artifacts. Every expression of design can potentially become an act of contestation and a site for agonism. Participatory practices, in particular, can leverage similar tactics of critical product design, such as ambiguity, and absurdity, to instantiate situations in which participants are challenged to make sense of counterintuitive activities as a way to promote critical engagement with a matter of public concern, object of a given participatory session (Lupetti et al., 2023).

STEERING STORIES

Steering Stories is a research project that applies **Adversarial Design** as a way to confront and challenge dominant narratives of driving automation (robot cars). In particular, the project revolves around the development of contestational artifacts as inquisitive tools to understand whether and how the discussions of strategic stakeholders do map to dominant discourses surrounding driving automation, as well as to promote the emergence of more nuanced storylines.

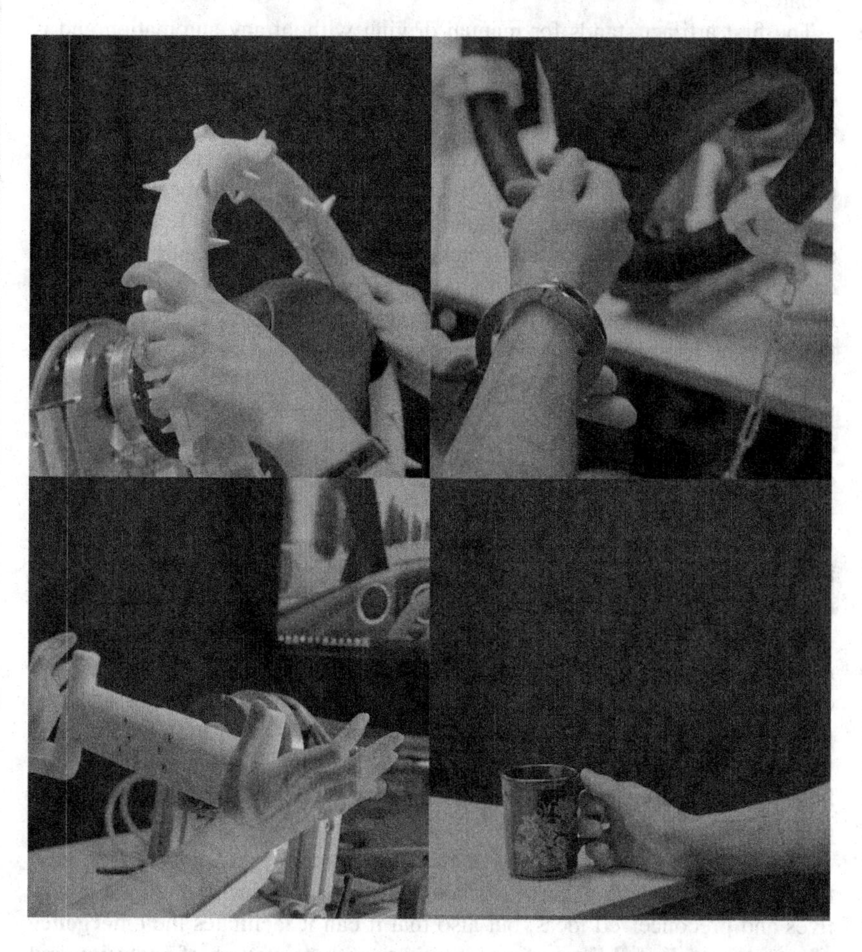

FIGURE 5.6 The four contestational steering wheels developed in the Steering Stories project. Top left: steering wheel with spikes representing full manual driving. Top right: steering wheel with handcuffs representing traded control driving. Bottom left: game-like handles to steer, representing shared control driving. Bottom right: no steering wheel representing full autonomous driving.

The contestational artifacts consist of a series of four provocative steering wheels, each characterized by a feature that generates cognitive estrangement and sustains ambiguity. Specifically, the artifacts were ideated by selecting an emblematic element of driving, the steering wheel, and redesigning it to suggest a specific narrative stand toward automation and the related implicit trade-offs between driving comfort and safety. This way, reductionist, and controversial ideas are manifested with the intent of confronting and provoking debate.

The first artifact stands for manual driving without any automation and is characterized by the addition of spikes, which are an allegory for the inherent effort and danger that a driver can experience. The second presents a perspective where control is never fully delegated to the car, rather constantly shared between the vehicle and the human driver. As the underlying idea is that automation is in support of the human driver, helping to cope with their limitations, the characterizing elements are arm supports and handles that remind of both a videogame-style interface and crutches. The third artifact stands for a traded control interaction strategy of automated vehicles, where either the human or the car is in control of the driving tasks. This aims to relieve the human from the effort of driving, and potentially to do something else in the meantime, yet the driver remains always responsible for the behavior of the car and should be ready to take back control at all times. To represent this controversy, a pair of handcuffs added to the steering wheel to make explicit the responsibility demand that traded control implies.

In the last artifact the steering wheel disappears as it represents a perspective that sees full automation as the best solution to the multitude of issues related to driving. The need for a control interface ceases to exist and in place of that, d a table-like surface is added.

Building on existing narrative research approaches where argumentative resources are used as interpretative aids and boundary objects for debate and deliberation, the contestational artifacts were employed in a series of focus groups with experts, from engineering researchers to innovation experts, and from municipality managers to road safety experts, to understand whether and how we can break free from dominant narratives when discussing the future of these technologies.

Insights from the focus groups revealed that not only Adversarial Design in the form of contestational artifacts can confront publics with dominant narratives and preconceived ideas, but also that it can it facilitates the **emergence of usually neglected lines of questioning**, such as matters of exclusion and geo-politics.

BOX 5.7 ETHNOGRAPHIC EXPERIENTIAL FUTURES

Michal Luria

Ethnographic Experiential Futures (EXF) is a design research method that initiates a discussion about possibilities for the future. It examines the consequences of potential scenarios and considers alternatives to the status quo. The method involves two stages: (1) Identifying current visions of the future within a particular community or group; and (2) using these visions as a foundation for creating tangible and rich probes of the future, which can be used for reflection. The EXF approach could be valuable to researchers interested in **revealing underlying tensions and dilemmas** within a particular design space, while promoting ethical consideration and change in a range of technological fronts.

EXF builds on participatory foresight and critical design research methods, with the goal of making a more diverse array of scenarios available for deliberation. While within Speculative Design and Design Fiction traditions, EXF is different in that the designer or researcher takes a secondary role. Instead of defining speculative probe themselves, EXF researchers heavily rely on ethnographic interviews. In the context of Human-Robot Interaction, these interviews may be conducted with HRI researchers, robotics industry workers, robot owners, or others.

Led by ethnography principles, EXF is best used for early and exploratory stages of design processes. Its workflow includes four essential parts:

Mapping: Inquiry about the group's images of the future within a particular topic, including expected, desired and feared scenarios (e.g., an inquiry about robotics industry workers' perspectives of the future of work with robots.)

Mediating: Translation of discussions and ideas about future visions into tangible probes or immersive experiences.

Mounting: Staging of the identified experiences to present back to interviewees or to a broader audience, with the aim to form a deeper discussion about the futures in question.

(Second) Mapping: Investigation of audience responses to the presented experiences.

LETTERS FROM THE FUTURE

Letters from the Future was a design research project that used EXF to explore possibilities for future social agents while considering their potential impacts and consequences (Luria & Candy, 2022). The focus was to learn about ideal visions of the future for social agents among researchers and designers in the field of Human-Robot and Human-Agent Interaction. Through their visions, this research identified ethical tensions that should be discussed, in parallel to the development of the technology.

FIGURE 5.7 Letter prototypes that expressed three different visions of future social agents by experts in the research and design fields. Participants received these letters as evocative probes to their home address.

For the mapping phase of EXF, in which data is collected to identify a series of specific images of the future by a group or community, three subject matter experts participated and shared their views of future social agents.

In the mediating stage, we analyzed ideas about ideal futures that surfaced in the interviews, and processed them into concrete form. The goal was to mediate each of the three conversations into a corresponding concept of a future social agent, one per interview.

For mounting, each defined concept was converted into a tangible "letter probe," designed to bring a "what if?" speculation about future social agents to life. 15 researchers in the fields of HRI and HAI were invited to experience these probes—they received three physical letters to their homes over the course of three weeks, each telling a story of a future social agent. The letters attempted to capture agent concepts beyond their technical capabilities, but about the world in which they exist. While experiential future probes can take many tangible forms, the use of paper letters had several advantages: They told a complex story while keeping prototyping to a minimum; they enabled inexpensive but high-fidelity provocations; and they created experiences remotely and asynchronously, bringing ideas to a broader audience who were not co-located.

In the final mapping stage, we collected participants' responses and reflections on the letter probes through open-ended surveys online. Through qualitative analysis, we identified themes about the community's ethical considerations and concerns with these anticipated futures.

The paper (Luria & Candy, 2022) discusses the deployment of EXF in the space of social robotics, considering both successes and weaknesses, and suggesting possible improvements for future work. It is a contribution towards better practices of research-through-design using prototyping and futures methodologies, and highlights the value that these approaches could bring into professional communities of technology researchers and designers.

BOX 5.8 BLENDING TRADITIONS

Haipeng Mi

Anthropologists regard tradition as "an ongoing cultural creation" that provides insights into community collective consciousness and intergenerational renewal. Gross (1992) delved deeply into the relationship between the continuity of tradition and modernity, hinting that our past practices and narratives directly shape current behaviors and mentalities. All these premises point to one question: As designers, engineers or any practitioners living in the present, how should we blend tradition and culture into robot interaction design—an activity full of modern technologies and knowledge—and generate new experiences, memories and traditions?

Firstly, we need to recognize the significant role of tradition in robot design. The aesthetic emotions and symbolic systems embodied in traditional culture provide abundant inspirations for robot appearance design, motion design and interaction design. The blending of traditional cultural elements makes robot interaction not only full of unique charm but also more likely to **evoke users' cultural identity and sense of belonging**, which is especially suitable for venues like museums and theme parks displaying traditional culture, while also providing new opportunities for preserving, disseminating, popularizing and even innovating traditional culture. Secondly, we can regard tradition as a "cultural coordinate", whose inherent etiquette and social systems will strongly constrain aspects of robots like language style, communication method and behavior norms, enabling robots to better fit into the broader, more abstract cultural context.

In design practice, we usually need to consider various factors such as aesthetic style, task scenario, interaction style, and even manufacturing technology and operation environment, ensuring these different elements maintain consistency, integrity, and adaptability in specific contexts to achieve optimal collaborative effects.

MOJA ROBOTIC CHINESE ORCHESTRA

The Moja robotic Chinese orchestra (Figure 5.8) represents a proactive attempt at combining Chinese traditional culture with robot design, infusing inspirations and creativity from tradition into robot design. This orchestra consists of three robot musicians: Yaoguang, Kaiyang and Yuheng, playing traditional Chinese folk instruments like konghou, set drums and bamboo flute, capable of collaborating with human performers to deliver high-quality concerts or stage performances.

FIGURE 5.8 Three robotic musicians of Moja (from left to right): Kaiyang, Yuheng, and Yaoguang.

Tradition and culture frame the top-level design context, posing significant constraints on robots' persona and behavior mechanisms. Born in Chinese traditional culture, Moja implies that the exterior and interaction styles of the robots should follow the inherent cognitive, aesthetic and social standards of Chinese tradition, representing a blending. For example, the appearances and names of the robot musicians draw from Chinese traditional cultural allusions, tracing the robots back to historical lineages.

We employ Chinese folk music as a non-verbal communication language, highlighting its unique playing methods and timbre as a highly condensed ethnic memory. More specifically, the shells of Moja robot musicians are made using traditional wood carving crafts; the natural material and quaint style contrast with the technological attribute of robots, endowing the robots with historical qualities and tender textures. It is worth mentioning that we designed multiple costumes for Moja robots catering to different performance scenarios, covering elegant, breezy and luxurious styles to fit the robots into diverse role settings in traditional contexts.

Meanwhile, we strived to incorporate traditional cultural narrative methods into performance design. Moja robots serve as both clues in the narrative and promoters of scene transitions and storyline progression, collaborating with human actors to present sophisticated performances on a limited stage. This minimalist narrative logic originates from traditional Chinese literature like novels, legends and oral traditions, which can inspire the understanding and imagination of the audience, representing an interaction principle embedded into Moja robots.

We transformed the ideas implied in tradition into two layers of stories: The richly plotted stage content and other metaphorical content, jointly constructing the delivery of traditional Chinese ideology.

The multiple performances of the Moja robotic Chinese orchestra have received positive feedback, proving our notion of adopting traditional culture as a design principle. It also provided possibilities and pathways for the continuation and innovation of traditional culture facing various challenges (Li et al., 2019).

REFERENCES

Alves-Oliveira, P., Lupetti, M. L., Luria, M., Löffler, D., Gamboa, M., Albaugh, L., Kamino, W., Ostrowski, A., Puljiz, D., Reynolds-Cuéllar, P., & Lockton, D. (2021, June). Collection of metaphors for human-robot interaction. In *Designing Interactive Systems Conference 2021* (pp. 1366–1379).

Auger, J. H. (2012). *Why Robot? Speculative design, the domestication of technology and the considered future*. Royal College of Art.

Auger, J. H. (2014). Living with robots: A speculative design approach. *Journal of Human-Robot Interaction, 3*(1), 20–42.

Babich, B. (2012). Martin Heidegger on Günther Anders and Technology: On Ray Kurzweil, Fritz Lang, and transhumanism. *Journal of the Hannah Arendt Center for Politics and Humanities at Bard College, 1*, 122–44.

Bardzell, J., & Bardzell, S. (2013). What is "critical" about critical design? In *Proceedings of the SIGCHI conference on human factors in computing systems* (pp. 3297–3306).

Candy, S., & Dunagan, J. (2017). Designing an experiential scenario: The people who vanished. *Futures, 86*, 136–153.

Carlyle, T. (2015). The mechanical age. *Industrialisation and Culture*, 1830–1914, 21.

Cave, S., & Dihal, K. (2020). The whiteness of AI. *Philosophy & Technology, 33*(4), 685–703.

Cheon, E., Sher, S. T. H., Sabanović, Š., & Su, N. M. (2019, June). I beg to differ: Soft conflicts in collaborative design using design fictions. In *Proceedings of the 2019 on designing interactive systems conference* (pp. 201–214).

Cheon, E., & Su, N. M. (2017, February). Configuring the user: "Robots have Needs Too". In *Proceedings of the 2017 ACM conference on computer supported cooperative work and social computing* (pp. 191–206).

Cheon, E., & Su, N. M. (2018, February). Futuristic autobiographies: Weaving participant narratives to elicit values around robots. In *Proceedings of the 2018 ACM/IEEE international conference on human-robot interaction* (pp. 388–397).

Cheon, E., Zaga, C., Lee, H. R., Lupetti, M. L., Dombrowski, L., & Jung, M. F. (2021, October). Human-machine partnerships in the future of work: Exploring the role of emerging technologies in future workplaces. In *Companion publication of the 2021 conference on computer supported cooperative work and social computing* (pp. 323–326).

DiSalvo, C. (2012). *Adversarial design*. MIT Press.

Dunne, A., & Raby, F. (2024). *Speculative Everything, With a new preface by the authors: Design, Fiction, and Social Dreaming*. MIT press.

Escobar, A. (2018). *Designs for the pluriverse: Radical interdependence, autonomy, and the making of worlds*. Duke University Press.

Fisher, W. R. (1985). The narrative paradigm: In the beginning. *The Journal of Communication, 35*(4), 74–89.

Fry, T. (1999). *A new design philosophy: An introduction to defuturing*. UNSW Press.

Gamboa, M., Baytaş, M. A., Hendriks, S., & Ljungblad, S. (2023, February). Wisp: Drones as Companions for Breathing. In *Proceedings of the seventeenth international conference on tangible, embedded, and embodied interaction* (pp. 1–16).

Gasparetto, A. (2016). Robots in history: Legends and prototypes from ancient times to the industrial revolution. In *Explorations in the History of Machines and Mechanisms: Proceedings of the Fifth IFToMM Symposium on the History of Machines and Mechanisms* (pp. 39–49). Springer International Publishing.

Gross, D. (1992). *The past in ruins: Tradition and the critique of modernity*. University of Massachusetts Press.

Hampton, G. J. (2015). *Imagining slaves and robots in literature, film, and popular culture: reinventing yesterday's slave with tomorrow's robot*. Lexington Books.

Höök, K. (2018). *Designing with the body: Somaesthetic interaction design*. MIT Press.

Huyssen, A. (1986). The vamp and the machine: Fritz Lang's metropolis. *After the great divide: Modernism, mass culture, postmodernism* (pp. 65–81). Palgrave Macmillan UK.

Jasanoff, S., & Kim, S. H. (Eds.). (2019). *Dreamscapes of modernity: Sociotechnical imaginaries and the fabrication of power*. University of Chicago Press.

Kozubaev, S., Elsden, C., Howell, N., Søndergaard, M. L. J., Merrill, N., Schulte, B., & Wong, R. Y. (2020, April). Expanding modes of reflection in design futuring. In *Proceedings of the 2020 CHI Conference on Human Factors in Computing Systems* (pp. 1–15).

La Delfa, J., Baytas, M. A., Patibanda, R., Ngari, H., Khot, R. A., & Mueller, F. F. (2020, April). Drone chi: Somaesthetic human-drone interaction. In *Proceedings of the 2020 CHI conference on human factors in computing systems* (pp. 1–13).

Li, J., Hu, T., Zhang, S., & Mi, H. (2019, June). Designing a musical robot for Chinese bamboo flute performance. In *Proceedings of the seventh international symposium of Chinese CHI* (pp. 117–120).

Lockton, D., Singh, D., Sabnis, S., Chou, M., Foley, S., & Pantoja, A. (2019). New metaphors: A workshop method for generating ideas and reframing problems in design and beyond. In *Proceedings of the 2019 on creativity and cognition* (pp. 319–332).

Lupetti, M. L., Zaga, C., & Cila, N. (2021, March). Designerly ways of knowing in HRI: Broadening the scope of design-oriented HRI through the concept of intermediate-level knowledge. In *Proceedings of the 2021 ACM/IEEE International Conference on Human-Robot Interaction* (pp. 389–398).

Lupetti, M. L., Cavalcante Siebert, L., & Abbink, D. (2023, April). Steering stories: Confronting narratives of driving automation through contestational artifacts. In *Proceedings of the 2023 CHI conference on human factors in computing systems* (pp. 1–20).

Luria, M., & Candy, S. (2022, April). Letters from the future: Exploring ethical dilemmas in the design of social agents. In *Proceedings of the 2022 CHI conference on human factors in computing systems* (pp. 1–13).

Luria, M., Sheriff, O., Boo, M., Forlizzi, J., & Zoran, A. (2020). Destruction, catharsis, and emotional release in human-robot interaction. *ACM Transactions on Human-Robot Interaction (THRI)*, *9*(4), 1–19.

Mayer-Schönberger, V., & Cukier, K. (2013). *Big data: A revolution that will transform how we live, work, and think*. Houghton Mifflin Harcourt.

Mazé, R. (2019). Politics of designing visions of the future. *Journal of Futures Studies: Epistemology, Methods, Applied and Alternative Futures*, *23*(3), 23–38.

Minden, M., & Bachmann, H. (2002). *Fritz Lang's Metropolis: Cinematic Visions of Technology and Fear*. Camden House.

Pelzer, P., & Versteeg, W. (2019). Imagination for change: The Post-Fossil City Contest. *Futures*, *108*, 12–26.

Rhee, J. (2018). *The Robotic Imaginary: The Human and the Price of Dehumanized Labor*. University of Minnesota Press.

Šabanović, S. (2010). Robots in society, society in robots. *Advanced Robotics: The International Journal of the Robotics Society of Japan*, *2*(4), 439–450.

Schaffer, S. (1994). Babbage's intelligence: Calculating engines and the factory system. *Critical Inquiry*, *21*(1), 203–227.

Silverstone, R. & Hirsch, E. (1992). *Consuming technologies: Media and information in domestic spaces*. Routledge.

Stephens, E., & Heffernan, T. (2016). We have always been robots: The history of robots and art. In D. Herath, C. Kroos, & Stelarc (Eds.), *Robots and art: Exploring an unlikely symbiosis* (pp. 29–45). Springer Singapore.

Tennent, P., Höök, K., Benford, S., Tsaknaki, V., Ståhl, A., Dauden Roquet, C., … & Zhou, F. (2021, May). Articulating soma experiences using trajectories. In *Proceedings of the 2021 CHI conference on human factors in computing systems* (pp. 1–16).

Trovato, G., De Saint Chamas, L., Nishimura, M., Paredes, R., Lucho, C., Huerta-Mercado, A., & Cuellar, F. (2021). Religion and robots: Towards the synthesis of two extremes. *Advanced Robotics: The International Journal of the Robotics Society of Japan*, *13*(4), 539–556.

Verbeek, P. P. (2006). Materializing morality: Design ethics and technological mediation. *Science, Technology, & Human Values*, *31*(3), 361–380.

Winkle, K., McMillan, D., Arnelid, M., Harrison, K., Balaam, M., Johnson, E., & Leite, I. (2023, March). Feminist human-robot interaction: Disentangling power, principles and practice for better, more ethical HRI. In *Proceedings of the 2023 ACM/IEEE international conference on human-robot interaction* (pp. 72–82).

Zaga, C., & Lupetti, M. L. (2022). Diversity equity and inclusion in embodied AI: Reflecting on and re-imagining our future with embodied AI. *research.utwente.nl*. https://research.utwente.nl/files/285680270/DEI4EAIBOOKLET_WEB_SINGLEPAGES.pdf

6 Choosing Materials for Personal Robot Design

Guy Hoffman
Cornell University, Ithaca, United States

Choosing materials for human-made objects has been at the center of a millennia-long tradition of artifact design and production. As humanity accelerated through a sequence of found and manufactured materials, designers increasingly had to carefully choose the appropriate material for the task at hand. Should a container be made of fired clay or woven reed fibers? Will a table be made of heavy oak wood, or a thin marble plate supported by wrought iron legs? Is aluminum required for this garden shed, or is PVC sufficient? A glance around the reader's desk will likely reveal a collection of human-made objects which incorporate a vast diversity of material choices echoing humanity's trajectory through its history, from wood, stone, and bamboo fibers, through metals, fabrics, paper, and ceramics, to synthesized plastics and engineered composites.

The design of personal robots stands in stark contrast to this rich tradition, with most robot designers making seemingly automatic choices, informed by convenience, functionality, and pragmatism. The result is a near consensus on plastics and metals as the materials used, and a concerning similarity in the design of robots that interact with humans. This lack of creativity could be excused in an industrial setting, where engineering considerations dominate, but should be questioned when it comes to the design of robotic devices that would make their way into our living, leisure, and educational spaces. There are, of course, good arguments to be made for pragmatic choices of convenient materials, including price, availability, ease of manufacturing, machinability, and strength. That said, the choice of materials offers designers modes of interaction and expression which are beyond mere functionality and should be incorporated into personal robot design.

I will suggest four ways in which materials matter for human–object interaction and which can be used to better inform the design of personal robots: visual aesthetics, tactile interaction, auditory expression, and a relationship to time and place.

6.1 VISUAL AESTHETICS

Most apparently, materials communicate to humans in visually aesthetic ways, which are in turn informed by cultural references. Physically, different materials capture and reflect light in unique ways, displaying specific colors and textures. Deep-shaded

DOI: 10.1201/9781003371021-6
This chapter has been made available under a CC-BY-NC-ND license.

wooden furniture, rough textured or glazed clay pots, and colorful stained-glass lampshades affect the observer from a distance, before direct interaction even starts. The emitted colors can be categorized as warm or cool, and the textures as inviting or unpleasant.

Beyond their physicality, materials can also visually radiate cultural histories and traditions. Any choice of material recalls thousands of years of craft skills, honed and reimagined by generations of designers and artisans. When one observes a manufactured object, one can see how generations of woodworkers, potters, metalsmiths, glass blowers, and weavers have embedded their ideas into the object. These ideas carry meaning: the difference between placing a rough-sawn wooden farm table in the center of a dining area and choosing a chromed steel and laminated diner-style table embeds a narrative, either of tradition and rural life or of modernity and convenience.

Even if ignored by the designer, the choice of usually white plastic and metal for personal robot design also carries visual aesthetic and cultural connotations (see: Dunstan & Hoffman, 2023). A designer might only think of functionality when choosing materials for a personal robot, but they might inadvertently tell a story of futurism and consumer convenience, while leaving a wide space of design potential unexplored.

6.2 TACTILE INTERACTION

Coming closer to the manufactured object, humans interact with artifacts by touching them. When touched, materials interact with the user's skin in rich ways that designers have long taken advantage of. Materials have unique textures and a characteristic resistance to deformation, which are quickly experienced by anyone who handles them. Running a hand over a glass lampshade shares very little with the experience of touching a fabric one. A vegetable peeler made purely of stainless steel performs differently in a cook's hand than one with engineered plastic handles.

In addition to texture, a material's thermal conductivity causes it to feel colder or warmer to the touch, as can be experienced by touching either metal or fleece, both at room temperature. While this property of materials is sometimes exploited by robot designers, for example by adding fur to robots that are expected to be held (e.g., Paro, the robotic baby seal; Wada & Shibata, 2007), most personal and social robots are cold to the touch.

What would it feel like to touch a wooden robot versus one made of marble? Would a robot covered in fleece be preferred in the winter, and a cool glass one serve better in the summer? The tactile expressivity of a robot's exterior offers a wide canvas for design exploration (see: Hu & Hoffman, 2023), one that is yet to be fully discovered.

6.3 AUDITORY EXPRESSION

Touching an object, picking it up, tapping it, stroking it, dropping it, placing it down—all of these interactions produce characteristic sounds that depend on a combination of material choice and shape, in interaction with the surface onto which the object may have been released. Setting down a thin vessel of glass makes a sound that reflects its fragility, running one's fingers through leaves of paper generates a rustle, and metal spoons resonate in idiosyncratic ways when they are dragged across a cast-iron pan or a glazed dinner plate.

While sound has been explored as a nonverbal modality in human–robot interaction (e.g., Moore et al., 2017; Pelikan & Jung, 2023; Wolfe et al., 2024), these works dealt mostly with computationally produced sounds or peripheral sounds, such as those emitted by electric motors and gear transmissions. The auditory expression of the materials chosen to construct the machine is rarely considered when designing a robot.

We can imagine a robot with a shell made of paper that makes crumpling, folding, and unfolding sounds as it moves. Or one made of pieces of fired clay that rub against each other, generating the scratching noises of erosion. Even when a mobile personal robot hits an obstacle, the material from which it is made can be designed to create a certain sound, be it a soft rubber thump, or an alarming shattering of glass.

6.4 TIME AND PLACE

Finally, material choice relates to the designed object beyond the immediate interaction with users. Each design material has a longitudinal temporal relationship to the world, which projects in both directions, into its history and its future. Looking to the past, materials are chosen for their location of origin, be it to support local production or remind one of a specific time or place in the world, from which a material originated. Many enjoy furniture made from reclaimed wood planks that still carry painted shipping stamps from bygone centuries. A set of coasters crocheted from yarn brought on a boat by an immigrant grandparent might remind one of their ancestry. Looking into the future, people may choose objects made from certain materials for their longevity, as well as for their environmental impact, preferring degradable materials to those who will pollute oceans for centuries.

Personal robot design rarely taps into its potential to connect to distant times and places. Robots come from nowhere, designed by no one, made from materials that don't have a spatiotemporal identity, whether in the past or in the future. To use the famous term coined by Benjamin (1936), they don't have a unique "presence in time and space." When they are done operating, they will disappear into a landfill, forgotten, with some notable exceptions like the mindful rituals surrounding the zoomorphic AIBO robot in Japan (Knox & Watanabe, 2018).

When choosing materials for a personal robot, designers should consider their origins and fate. If long-term interaction is a goal, one can think longer-term than just the period of the robot's usage.

The four interactive considerations in material choice listed above come together to produce a holistic affective relationship between humans and their artifacts, a relationship completely missed by designers of social and personal robots, even as they proclaim to design robots that are intended for long-term, personal, and emotional interaction. It should therefore not be surprising if users are more attached to their hand-made clay tea mug than to their personal robotic assistant.

How can the field of personal robot design address this blind spot in material choice? One way is to formalize material choice in the education process of designers in the human–robot interaction field. Product designers and architects are meticulously trained in material selection from not just a pragmatic, but also a cultural and affective perspective. When they begin to practice, they carry this training and sensitivity into their trade. Designers of social robots should take similar care in the materials they offer their users.

6.5 DESIGNING ROBOTS WITH AN INTENTIONAL MATERIAL CHOICE: BLOSSOM

I would like to offer *Blossom*, in Figure 6.1, as a case study of intentional material choice in a robot's design (Suguitan & Hoffman, 2019). Blossom was designed in my research laboratory as a speculative project reacting to the common tropes in personal robot design, especially that of a streamlined white plastic appliance. One design principle of the Blossom project was the use of traditional craft materials, most prominently wool and wood.

The choice of craft materials was made with several objectives in mind. First, aesthetically, I wanted to position a personal robot in the material context of domestic objects. A cold consumer electronics device feels out of place on a hand-made wooden coffee table. Second, Blossom was made compliant and out of a soft crocheted fabric to encourage tactile interaction, including holding, petting, pushing, and squeezing the robot. Third, the choice of craft was to encourage amateur manufacturing and social interaction. I envisioned a grandparent and grandchild, neither of which might have access to a metal shop or a 3D printer, working together to build a Blossom robot (see the interior and components in Figure 6.2 and 6.3). Finally, the natural materials chosen for Blossom would eventually decay, contributing both to a reduction in waste-related pollution, and lending a temporary aspect to the ownership of the robot (for a discussion on the value of transience in social robot design, see: Hoffman, 2020).

Blossom was adopted by other researchers as a low-cost hand-made social robot (e.g., Swaminathan et al., 2021; O'Connell et al., 2024). Its adoption was perhaps due to the robot's accessibility to production or because of the joy of crafting one's own robot. Still, Blossom is merely a single speculative suggestion that can expand the scope of material imagination in personal robot design.

To productively move forward as a field toward intentional material choice, we should encourage collaborations between robotics engineers and traditionally trained product designers. The latter would bring a more nuanced, aesthetic, and culturally grounded approach to material choice in robot design. Robot designers themselves would also benefit from both a more elaborate product design education with a focus on material choice, which may result in a more intentional process when choosing materials for their next project.

FIGURE 6.1 The Blossom robot displaying a hand-crocheted wool cover and handcrafted wooden ears. These materials were intended to aesthetically fit in domestic environments, be warm to the touch, and evoke a sense of a unique time and place.

Photo courtesy of Cornell University.

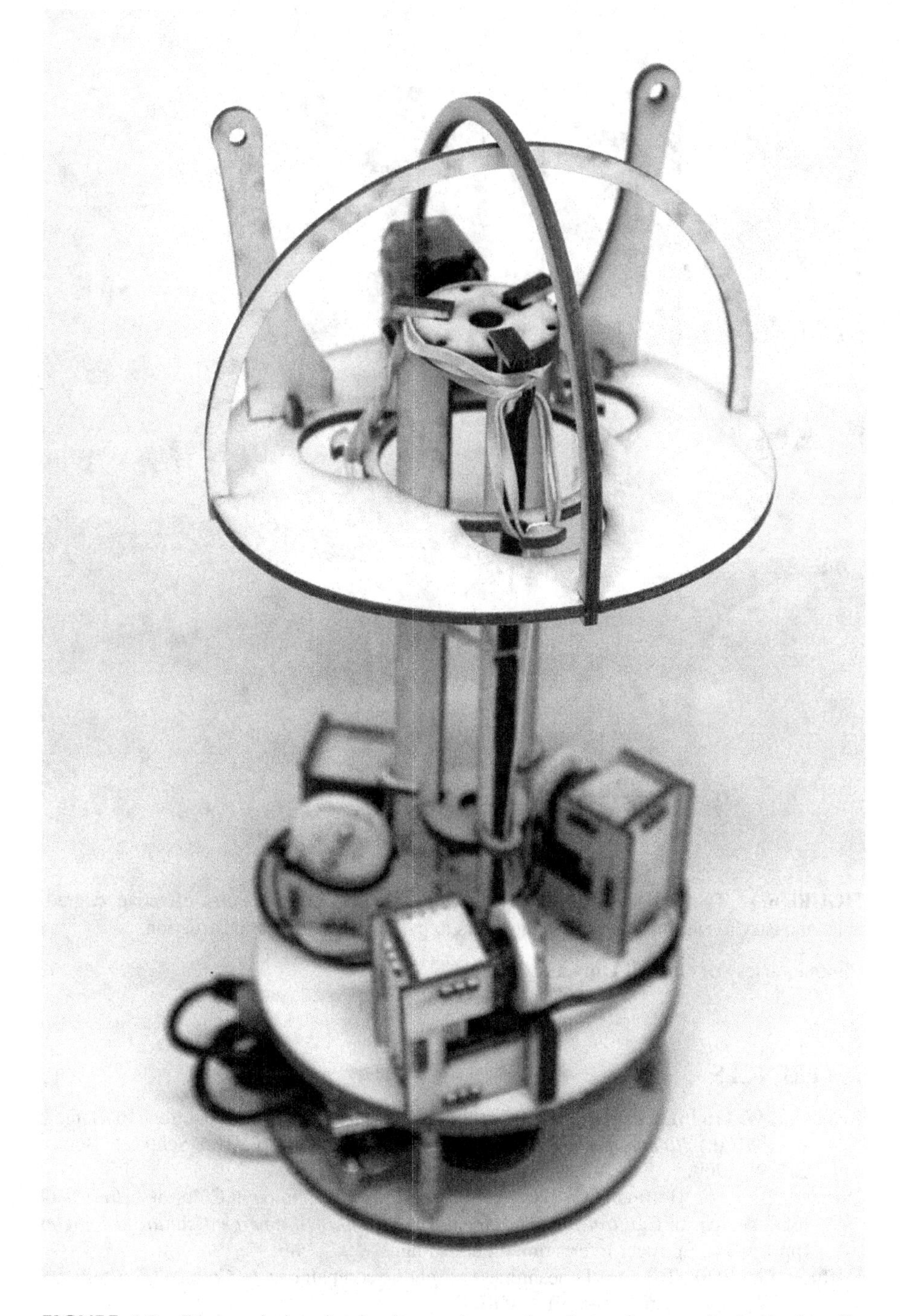

FIGURE 6.2 Blossom's interior is also made mostly of wood, an ecologically friendly, degradable material, that suggests temporality and accessibility.

Photo courtesy of Cornell University.

FIGURE 6.3 The wooden mechanical parts Blossom is made of are cost-effective, degradable, and easy to assemble, enabling amateurs to participate in robot construction.

Photo courtesy of Cornell University.

REFERENCES

Benjamin, W. (1936). The work of art in the age of mechanical reproduction. In Hannah Arendt (Ed.), *Illuminations: Essays and reflections*, trans. Harry Zohn. Schocken Books, 217:52, 1969.

Dunstan, B. J., & Hoffman, G. (2023). Social Robot Morphology: Cultural histories of robot design. In *Cultural robotics: Social robots and their emergent cultural ecologies* (pp. 13–34). Springer International Publishing.

Hoffman, G. (2020). The social uncanniness of robotic companions. In *Culturally sustainable social robotics* (pp. 535–539). SAGE.

Hu, Y., & Hoffman, G. (2023). What can a robot's skin be? Designing texture-changing skin for human–robot social interaction. *ACM Transactions on Human-Robot Interaction, 12*(2), 19 pages.

Knox, E., & Watanabe, K. (2018). AIBO robot mortuary rites in the Japanese cultural context. In *2018 IEEE/RSJ international conference on intelligent robots and systems (IROS)* (pp. 2020–2025). IEEE.

Moore, D., Tennent, H., Martelaro, N., & Ju, W. (2017). Making noise intentional: A study of servo sound perception. In *Proceedings of the 2017 ACM/IEEE international conference on human-robot interaction (HRI '17)* (pp. 12–21). Association for Computing Machinery.

O'Connell, A., Banga, A., Ayissi, J., Yaminrafie, N., Ko, E., Le, A., Cislowski, B., & Mataric, M. (2024). Design and evaluation of a socially assistive robot schoolwork companion for college students with ADHD. In *Proceedings of the 2024 ACM/IEEE international conference on human-robot interaction (HRI '24)* (pp. 533–541). Association for Computing Machinery, New York, NY.

Pelikan, H., & Jung, M. (2023). Designing robot sound-in-interaction: The case of autonomous public transport shuttle buses. In *Proceedings of the 2023 ACM/IEEE international conference on human-robot interaction (HRI '23)* (pp. 172–182). Association for Computing Machinery, New York, NY.

Suguitan, M., & Hoffman, G. (2019). Blossom: A handcrafted open-source robot. *ACM Transactions on Human-Robot Interaction (THRI)*, *8*(1), 1–27.

Swaminathan, J., Akintoye, J., Fraune, M. R., & Knight, H. (2021). Robots that run their own human experiments: Exploring relational humor with multi-robot comedy. In *2021 30th IEEE international conference on robot & human interactive communication (RO-MAN)* (pp. 1262–1268), Vancouver, BC.

Wada, K., & Shibata, T. (2007). Living with seal robots—Its sociopsychological and physiological influences on the elderly at a care house. *IEEE Transactions on Robotics*, *23*(5), 972–980.

Wolfe, H., Su, Y., & Wang, J. (2024). Dimensional design of emotive sounds for robots. In *Proceedings of the 2024 ACM/IEEE international conference on human-robot interaction (HRI '24)* (pp. 791–799). Association for Computing Machinery, New York, NY.

7 Designing Robots that Work and Matter

Carla Diana
Cranbrook Academy of Art, Bloomfield Hills, United States

It started with ordinary things: coffee machines, printers, handheld thermometers, and the like. As a designer out of school, I always knew that I wanted to work on physical products with behaviors enabled by digital components, and had the good fortune of joining highly regarded firms, including Smart Design, frog Design, and Karim Rashid's Studio. Though I'd dreamed of the chance to design something as fantastical and sci-fi visionary as a real, live robot, it never seemed like it could become a real opportunity. Years later, when working as a Visiting Assistant Professor at Georgia Tech, an email appeared in my inbox announcing the search for an industrial designer to join a robotics lab. I jumped at the chance, got the gig, and was brought on as part of the core team of a project for the Institute's newly formed Socially Intelligent Machines Lab, led by Dr. Andrea Thomaz.

At the time, I had never been exposed to social robotics, a field that focuses on interacting with computing machines through intuitive, social-based behavior. I was fascinated and immediately hooked, beginning a career-changing trajectory into a specialty of designing for social robots. The Socially Intelligent Machines (SIM) Lab was focused on enabling robots to function in human environments "by allowing them to flexibly adapt their skill set via learning interactions with end-users." The lab focused on Socially Guided Machine Learning (SG-ML), "exploring how Machine Learning agents can exploit principles of human social learning." The result was an expressive upper torso humanoid robot platform called Simon that could be trained to perform tasks in various real-world situations, from clearing the table to sorting objects by color. It had eyelids that blinked, irises that rotated inside an eye socket, ears that tilted and twisted, and a head that rotated and nodded. Its torso could bend forward, and its shoulders, elbows, and five fully articulated fingers could be on each hand.

7.1 FROM THE LAB TO THE REAL WORLD

Fast forward 15 years to 2023, and I found myself walking down the hallway of my local hospital in suburban Detroit, closely following a sleek, cute, 4-foot-tall mobile robot named Moxi as it shuttled medication, supplies, lab samples, and personal items through hospital hallways, and traveled from floor to floor. My eight-year-old son, Massimo, was at my side, asking me 1,000 questions about what the robot was doing and how it worked. Moxi is a project whose design efforts I led for a

DOI: 10.1201/9781003371021-7
This chapter has been made available under a CC-BY-NC-ND license.

company called Diligent Robotics, cofounded in 2017 by Andrea Thomaz and Vivian Chu and made possible by a fantastic team of engineers and software developers. I was brought on board at the start of the company to help design a new robot platform based on the insights gleaned from observations in hospital settings as well as the deep learning in human–robot interaction (HRI) that emerged from the years of research at the SIM Lab. We knew that we wanted to capitalize on the kind of rapport humans and machines could have with each other and implemented a design story of the machine as a live, responsive, personality-driven entity. But we also wanted to create a device that was as minimal, unobtrusive, and practical as possible, to suit the efficient and hectic environment of the hospital setting. The result of years of research, collaboration, prototyping, and iteration was now in an actual work context.

Moxi has a minimal architecture that includes a head, a torso, and a storage base (see Figure 7.1). Besides physical gestures, it provides feedback via LED "eyes" a lighted headband, and an embedded touch screen. It has a robotic arm and a set of wheels on its base and can be programmed to run errands around the hospital. Nurses can set up rules and tasks when certain things change in a patient's record. For instance, if a patient has been discharged and their room is marked clean, Moxi will get a command to deliver an admission bucket (a set of fresh supplies for a new patient) so that it's ready for the next person. Nurses can also summon the robot to deliver or pick up items like lab tests and pharmaceuticals.

Seeing Moxi in action on site was certainly a thrill, but the most fascinating part of the visit was watching real people at their jobs interacting with it on an ordinary day. Massimo and I hadn't announced who we were, and the person who checked us in didn't alert anyone to our presence, so the view I got was a pure "fly on the wall" vantage point where I could observe the robot and the nurses working together on real tasks. It was amazing to see it integrated into the hospital environment as well as the social fabric of the workplace, and this satisfying moment confirmed that the design work was worthwhile and that the seemingly esoteric, big, academic open-ended questions, like "What if we could interact socially with a computing machine?" can wind up helpful in the real world.

7.2 THE EMERGENCE OF THE ROBOT–WORKER RELATIONSHIP

"A hospital introduced a robot to help nurses. They didn't expect it to be so popular," read the headline in *Fast Company*'s article by Katherine Schwab on July 8, 2019. "During the trials, Thomaz reports that the nurses and hospital staff had a similarly positive reaction–even from early skeptics. Some nurses were like, 'It creeps me out a little, I don't like robots, I'm not into AI,' Thomaz says. But by the end they [were] like, 'Hey Moxi, hey girl, how's it going?' It was dramatic, in a matter of two to three weeks." A similar *Wired* piece by Khari Johnson piece entitled, "Hospital Robots Are Helping Combat a Wave of Nurse Burnout," April 19, 2022, reported, "After two years of battling COVID-19 and related burnout, nurses say it's been a welcome relief. Nurses can hail Moxi robots from kiosks at nursing stations or send the robot a task via text message. Moxi might be used to transport items that are too big to fit into a tube system, like IV pumps; lab samples and other fragile cargo; or specialty items, like a slice of birthday cake." Massimo and I were there that day to satisfy

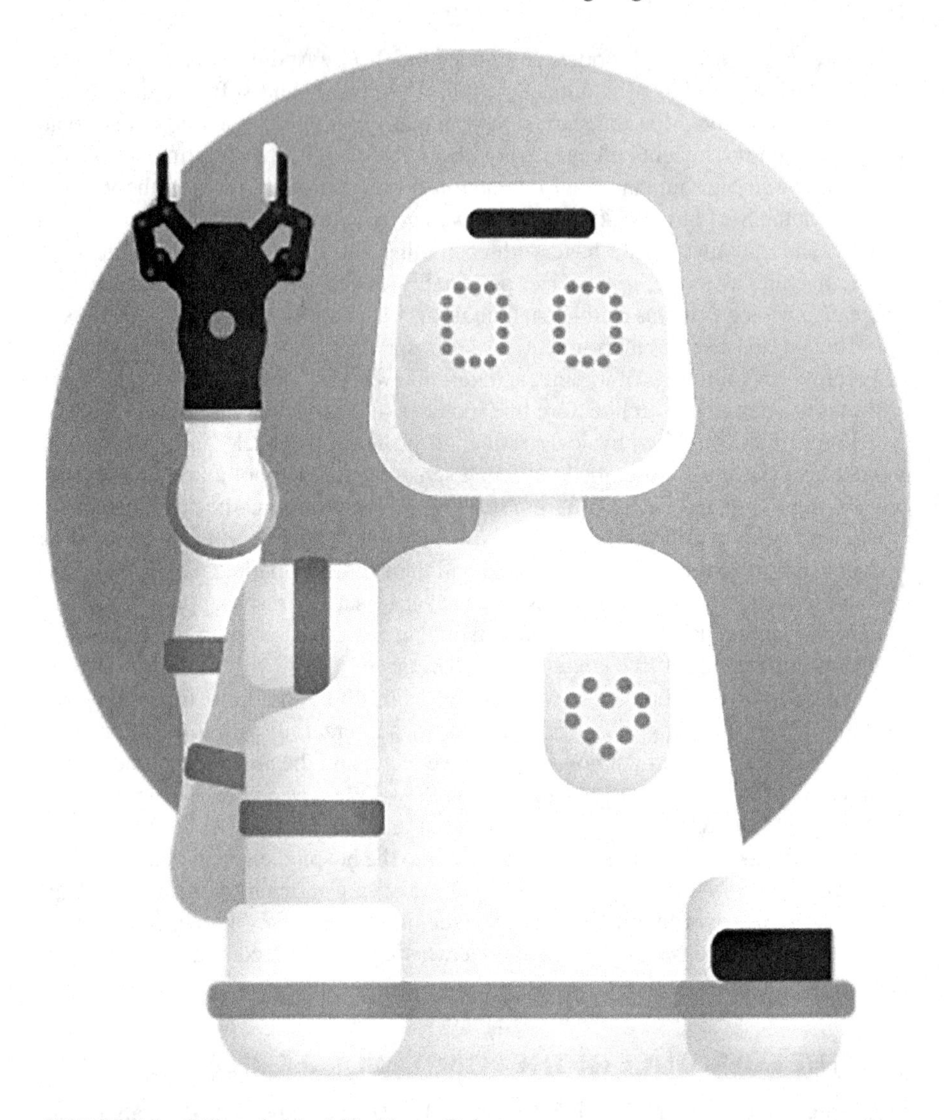

FIGURE 7.1 Diligent's Moxi Healthcare Robot.

Reprinted by permission of Harvard Business Review Press from MY ROBOT GETS ME: How Social Design Can Make New Products More Human by Carla Diana. **Copyright 2021 Carla Diana. All rights reserved.**

his curiosity. He's always looked over my shoulder at my sketchbook and computer screen and asked about my work, so this was my chance to give him a firsthand glimpse at the end result of some of that I do.

Moxi isn't designed to act like a nurse. Instead, the robot has been created to run the multitude of tasks nurses do that don't involve interacting with patients, like running errands around the floor or dropping off specimens for analysis at a lab. At the founding of Diligent Robotics, the product's core value was based on the observation that almost a third of a nurse's average day is spent on non-patient care: fetching, gathering, even taking out the trash. An American Nurses Association Study reported that a majority of nurses said Moxi gave them more time to talk with patients being discharged from the hospital, saved them energy, brought joy to patients and their families, and ensured that patients always had water when it was time for them to take their medication. Since Moxi's launch, the demand has continued to grow. In 2019, it was recognized as one of *TIME* Magazine's "Best Inventions" of the year.

During the process of designing the robot, I felt pulled between the poles of designing something that followed the austere trends of medical device design, and something that captured the animated nature of the robots we'd created in the lab. As a design philosophy in my own practice, I want to avoid the gratuitous use of humanoid characteristics. A great design should be as minimal as possible, using form to express only the most important aspects of interaction. I knew that our previous academic lab research included pushing the features of the robot to an almost exaggerated version of what the ultimate interface could be, yet I knew this was not what would be appropriate in the real world, especially the hectic and harried environment of the hospital hallway. When designing the robot, I continuously reminded myself that the social aspect is the true value.

7.3 KEY PRINCIPLES FOR ROBOT DESIGN

Reflecting on the process of designing Moxi, there are three interactive product design principles that I continue to keep in mind:

1. *Rely on in-depth design research with real people to gather true insights.* From my days in consultancies like Smart Design, I learned how valuable the simple act of talking to people can be. The heart and soul of the design process is research, so at the start of the Moxi project, we arranged several key interviews, giving us the opportunity to talk to stakeholders and spend time around real people on the job every day. These interviews helped in considering all aspects of the context, such as the demands of the physical environment, the experience of managing other equipment nearby, the state of mind of everyone involved, and even the impact of their clothing–if they are wearing gloves, for example, then aspects of the interaction would need to take that into account as a condition. Understanding context requires taking the time to truly understand how people will use products through design research. Initial design research allowed us to develop scenarios, or mental models of specific situations involving the person and the robot, that then served as a fundamental structure for imagining design needs.

   Learn as much as possible from quick-and-dirty mockups and design experiments. At the start of the design process for Moxi, we did a good deal of bodystorming–or embodied brainstorming–where people were asked to perform the actions that a person working with the robot would do. Using props made of simple, inexpensive, and malleable materials, we tried to simulate key situations as much as possible. For Moxi, we set up a mock hospital environment, using office furniture to approximate the shelving in the store room and a combination of plain boxes and actual medical products like syringes and gauze pads to use during fetch and delivery exercises. One person then played the role of a nurse, and someone else pretended to be the robot, equipped with a foam core screen with post-its for changing the on-screen messaging and for drawing different LED eye displays. We also had colored paper to indicate expressive lighting, or what eventually became an illuminated headband that could be seen at a distance to gauge the robot's status. With this basic setup, we could run through several typical scenarios, such as having a nurse summon the robot to deliver a welcome kit to a new patient or having it collect and deliver specimens to a lab. In each case, we enlisted other people as necessary to play the roles of passersby, hospital personnel, or patients. By re-creating the environment, we were able to glean important cues about the context of interaction that would otherwise not be apparent, and would likely not have been revealed through more removed representations such as drawings, renderings, or scaled models.

For example, by bodystorming with a mock environment, we got our first glimpses of aspects of the robot's context that would change how it might navigate doorways. We considered door hardware changes and ramp modifications, but also how the help of a nearby human might sometimes make the most sense. Working in life-size scale and in spaces that represented the final use offered deep and immediate insight into the situational constraints and opportunities. This technique allowed us to try out interactions in real time and space without the limitations or challenges of a prototype that had already been formed or constrained. It enabled a focus on the dialogue between the person and the robot while leaving the implementation details open for exploration and interpretation. We found ourselves thinking about whole body interactions, and we were less likely to think in terms of the more limited types of input architectures that have existed in products in the past. And by taking copious photos and notes with direct quotes and observations about context (time of day, nearby people, objects, fixtures, and positions), the documentation was later able to be synthesized to show an ideal workflow in the form of annotated scenario sketches.

2. *Continuously keep the team aligned.* In my practice, I've learned that keeping every aspect of an interactive product in mind is challenging, and this is especially important with something as complex as a robot. Bodystorming and research are a strong start, but they are only the beginning of a long process that can extend for months or years. As the design process advances into more concrete stages of design execution, a key challenge is keeping various team members, who all approach a project from their respective disciplines, aligned with the project priorities. For this reason I created the framework described in my book, *My Robot Gets Me: How Social Design Can Make New Products More Human*, as a way to guide team discussions around the various and simultaneous aspects of product performance taking place.

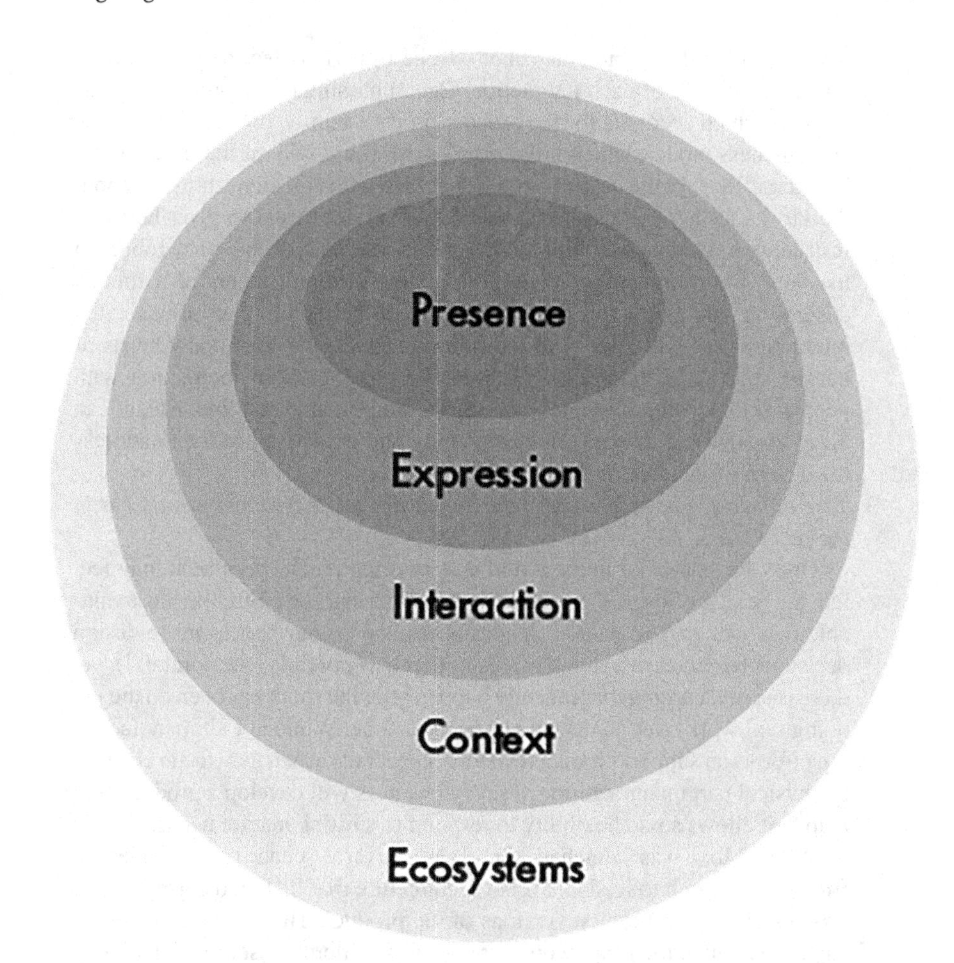

FIGURE 7.2 The My Robot Gets Me 5-part Framework for holistic project work alignment.

Reprinted by permission of Harvard Business Review Press from MY ROBOT GETS ME: How Social Design Can Make New Products More Human by Carla Diana. **Copyright 2021 Carla Diana. All rights reserved.**

It starts at the core, examining the **presence** of socially interactive products themselves. This aspect looks at the semantics implied by the forms and materials that might become part of the robot's structure, and how we might viscerally relate to them. In the case of a robot, this translates to embodiment and the overall anatomy as the core architecture. Next is a look at **expression**: what kinds of messages are communicated (e.g. "I've completed a task," "I'm running out of battery life"), how they are communicated, and to what effect. **Interaction** goes beyond expression to think of the back and forth dialogue which occurs when an object can sense and respond to people. The types of interactions we design and what they mean depend a lot on the **context** in which they take place; this includes the environment in which the robot will be used, but also the task, timing, purpose, and role of each interaction. Encompassing everything is the **ecosystem**, which accounts for the broader product family, product ecosystem, and business model, which influences the how and why of the product and its interactions. In the case of a robot like Moxi, this includes thinking about the kiosks at the nurse stations, the mobile devices used to give the robot instructions, storage bins and charging stations, and all the other robots in the fleet. By continuously examining all the different layers of a robot's design, we can gain a big picture view of all the factors that influence the interactions we will have with the future product, and how many levels of design need to be addressed to create successful social interaction with people. By focusing on scales of concern rather than discipline, a team can bring together the disparate areas of study and expertise needed to address the design of interactive products at each level.

3. *Take advantage of real-world deployment to gather feedback as it's out in the real world.*

    Once a robot is out in the world and working among people, it may feel like the design effort is done, but this is just the start of where the most valuable feedback can be gained. As designers, we do our best to make design decisions based on research that's as in-depth as possible (see point #1!), yet there is some learning that can only happen after the robot has been on the job doing real work over an extended period of time. While not all manufacturing efforts can support product revisions, especially when they mean changes to physical part manufacturing, a savvy business will develop a product road map that allows some flexibility to respond to shifting market needs.

    When Moxi was launched, there were several scenarios that emerged from our research to lead to a robot architecture that included a gripper that was installed with the first versions of the product. The use cases included situations where the robot would grasp and sort items, assemble bundles of products such as IV kits, manipulate environmental features such as door handles, and place items into its own drawer or external structures like baskets or shelves. Over time, it was discovered that a more valuable feature for making the robot as autonomous and efficient as possible would be the ability to accurately and reliably press elevator and remote door buttons so that it could swiftly navigate the hospital environment without needing human assistance. As there is a human at either end of the workflow for deliveries (a pharmacist to place medications in the drawer, for example, and a nurse to receive it on the other end), the manipulation we originally imagined was less important, but the robot's independence was a game-changer. Because of this discovery, we redesigned the robot to give it a button-pushing "hand" with smooth nubs approximating an index finger and thumb. Had we not taken advantage of the learning that can happen in the field over an extended period of use, we would have missed the opportunity to iterate on a key product feature. Similar iterations were able to take place for other features, such as refining the drawer configurations to adapt to specific needs for the items being transported.

7.4 ROBOT DESIGN IS EXTREME INTERACTION DESIGN

When I worked on the Simon robot in the research lab at Georgia Tech at the very start of my robot design career, it seemed like a project that was an exercise in exaggerated interface design. With its mesmerizing eyes, expressive ears, and highly gestural limbs, it represented the extreme of detailed product interaction. It was immensely exciting from a research perspective, but as a designer accustomed to the constraints of consumer mass market product design, it was challenging to envision a social robot whose benefit would justify the cost for such sophisticated technology. Now that Moxi is out in the world working with real people just up the street from me on an everyday basis, I see how the decades of passionate research in social robotics have leaned itself to focused, team-based real-world design iteration that can bring value of this research to everyday life in a way that truly helps people at work.

REFERENCE

Diana, C. (2021). *My robot gets me: How social design can make new products more human.* Harvard Business Press.

8 Critical Perspectives in Human–Robot Interaction Design

Sara Ljungblad and Mafalda Gamboa
Chalmers University of Technology, Gothenburg, Sweden

Robots depicted in science fiction and popular culture are typically imaginary creations with speculative features and unrealistic functionality. While these can aid us with future interpretations and the ethical deliberations of what robots can or should become, they also contribute to the shaping of robots, myths, and magical thinking. Such myths affect not only popular beliefs among the general public, but also how scholars in human–robot interaction (HRI) conduct research (see e.g. Fernaeus et al., 2009; Richardson, 2015; Sabanović, 2010). For instance, robot researchers may propose that robots will successfully supplant traditional human roles in everything from nursing to driving in heavy traffic.

In recent years, an underrepresented yet growing body of research in HRI has been aimed at addressing the need for more nuanced, situated, and critical perspectives in research. This includes calls to the research community with regard to exploring alternative epistemological groundings and associated critical topics in HRI (see e.g. Serholt et al., 2022, Ljungblad et al., 2018, Fernaeus et al., 2009, Bischof et al., 2022). The greater aim is to reframe HRI research toward designerly approaches that are more situated, reflective, critical, and inclusive. Examples of such endeavors could be to situate one's research in connection with official policies or guidelines, such as the ACM code of conduct (ACM code of conduct, 2023), or orient research with urgent topics proposed by (United Nations, 2023), which we explored in our first initiative at the workshop "Critical Robotics exploring a new paradigm" (Ljungblad et al., 2018). That time we used a backcasting method to connect emerging robot research themes and topics to UN goals. Since then, we have invited other researchers to contribute to a special issue with their perspectives of what constitutes critical robotics research (Serholt et al., 2022). The contributions highlighted aspects that are hidden or overlooked in research, providing many insights into the need for critical robotic perspectives and approaches. Several aspects related to methodology address the risks of oversimplification and deconstruction of work practices to fit a predefined technical solution (Burema, 2022; Maibaum et al., 2022) and acknowledge that user-centered methods can still lead to undesirable solutions (Dobrosovestnova et al., 2022; Weiss & Spiel, 2022). Another related issue was to reconsider design goals with human values at the center, for example, to create robots that manifest the value of reciprocity between humans rather than fake reciprocity in a social robot (van Wynsberghe, 2022).

DOI: 10.1201/9781003371021-8
This chapter has been made available under a CC-BY-NC-ND license.

In general, any research builds on moral or political values (Friedman, 1997), which raises questions about what constitutes genuine matters of concern within a research field (de La Bellacasa, 2011; Latour, 2008). Facts in research are not necessarily matters of concern in society as a whole, even when researchers stress that their research matters (Latour, 2008). If we develop more sensible ways to understand and successfully address real matters of concern in society, this may also lead to interesting discourses in research. This in turn could raise deeper critical questions about the design and use of robots in society as lenses into a bigger phenomenon.

> As artists continue to push the very limits of art, traditionally defined by discrete and inert handmade objects, they introduce robotics as a new medium and material at the same time that they challenge our understanding of robots - questioning therefore our premises in conceiving, building, and employing these electronic creatures.
>
> (Kac, 1997)

Compared to the myths we construct from science fiction movies, robotic art objects can be grounded in realistic technical functionality, and at the same time be norm-creative and innovative in their critical inquiry. Criticality in robotics can also successfully draw inspiration from other research fields, such as art, where the critical reconfiguration of robotics has essentially been done for more than 100 years (Csikszentmihalyi, 2022; Kac, 1997). Artistic methods can be used to question and explore our relationship to robots and robotic materials (Yolgormez & Thibodeau, 2022). For example, artists may create technically working solutions that hybridize robotic functionality with other media, systems, contexts, and even life forms (Kac, 1997). While doing so, they challenge our understanding of robots as a design material (Csikszentmihalyi, 2022; Jacobsson et al., 2013). This approach supports playful and critical reflection on robotics with the possibility of generating insights relevant for, but also far beyond social, aesthetic, and practical perspectives.

Verbeek (2011) describes how ethics and technology are deeply interwoven, and the need to make the implicit morality of things explicit in technology and other artifacts. In Human–Computer Interaction (HCI), several related critical research directions have surfaced through the years. Friedman (1997) contributed with the value-sensitive design perspective in 1997, articulating value and bias, and how technological artifacts inherently are political constructions. This in turn articulated the need for critical technical practice (Agre & Agre, 1997). Later on, perspectives and design approaches such as critical design (Dunne & Raby, 2001), reflective HCI (Dourish et al., 2004), criticality (J. Bardzell & Bardzell, 2013; Pierce et al., 2015), and adversarial design (DiSalvo, 2015) have emerged. Critical design (Dunne & Raby, 2001) addressed the everyday complexities of human pleasures, along with the misuse and abuse of electronic objects, and how complex realities challenge the promise of techno-utopian visions (J. Bardzell & Bardzell, 2013). Speculative design and adversarial design (DiSalvo, 2015) offer an additional lens on criticality approaches, such as prototyping a potentially evil design of an interface to understand ethical risks.

Speculative design and design fiction focus on producing illustrative scenarios, employing storytelling in the construction of imagined and hypothetical futures. Similar to artworks, critical and speculative design can push boundaries, challenge assumptions, and promote critical engagement with robot designs. Such work can go deeper than producing merely affirmative design. Without a critical stand, affirmative research may reinforce existing and occasionally problematic norms. Critical research instead engages directly with deterministic technological views; economic expectations; and oppressive structures. Such deeper and broader reflection can highlight and ground essential critical sensibility within the field (J. Bardzell & Bardzell, 2013). For example, Auger (2014) speculates and illustrates some robot-friendly aesthetics of existing artifacts in the home, such as plates and bed sheets. This supports conceptualizing robotic technical limitations and unexpected meanings such artifacts may create. Luria et al. (2020) focused on a non-utilitarian view of robots, creating speculations of destruction. Such adversarial design can shine light on misuse, and raise ethical concerns through fiction (DiSalvo, 2015). Parallel to these approaches, the practice of "undesigning" began to surface in the wider conversations associated with HCI (Pierce, 2012). Undesigning refers to the practice of reflecting on inaction as a possible activism, and the need to sometimes remove, exclude, replace, or restore technology. This recognizes that people may need something else than what they intentionally articulate, that some groups may be excluded from use, and how other groups may actively avoid use (Pierce, 2012, 2014). We can see the need for this when the implications of collected data clearly do not seem to support the initial design idea (e.g. a robot) (Baumer & Silberman, 2011).

Social science and humanistic research fields have a strong tradition of criticality, and we would like to see this embedded into the newer discipline of HRI. There is much to be contemplated after all – social structures, agency, power, and social dynamics. Such critical theory can also be profitably explored in conjunction with additional theoretical perspectives. We would benefit from an HRI discipline that was deepened through engagement with critical theory, as well as feminist theories (Gemeinboeck & Saunders, 2023; Winkle et al., 2023).

There is recognition of the need for this in the wider context of society. There are several overarching manifestos, projects, and organizations that point toward the harmful impact that AI and related technologies may have on society. They argue the need for ethical design and legislation. We can draw on these for inspiration. Examples include the Foundation of Responsible Robotics (a non-profit and non-government organization); Responsible AI (Dignum, 2019); and Trustworthy Robotics (Brando et al., 2022). Similarly, the Vienna Manifesto aims to build a community of scholars; policymakers; industry; and other actors to ensure that technological development remains centered on human interests (van Wynsberghe & Sharkey, 2020). To mitigate algorithmic bias, Black in Robotics (BiR) constitutes a community of Black researchers, industry professionals, and students with the joint aim of advocating for equity and ethical and equal outcomes.

8.1 CRITICAL ROBOTICS AS A RESEARCH PROGRAM

In this chapter, we propose Critical Robotics as a design research program. This is intended to support researchers in applying more critical and interdisciplinary perspectives in their work. We present this as a transitional theory – a conceptual structure aiming to support the formation of alternative research approaches and strengthen alternative epistemological grounds for HRI research. Research can navigate the design spaces between theory and practice; between art and science; and between freedom and method (Redstrom, 2017). Importantly, such approaches allow us to break away from the easy answers associated with our immediate intellectual habits. One intent of this is to investigate basic definitions of core concepts in design, allowing us to make and create them from new angles (Redstrom, 2017). A program is a way of dealing with complexity, in design and we do this by creating a composite definition of what designing is:

> In particular, programs allow us to work with matters pertaining to worldviews, the basic set of belief that design depends on, which are rarely made explicit in practice.
>
> (Redstrom, 2017)

According to Redstrom (2017), programs can perform several roles. A program connects basic definitions to methodology, for example, by illustrating what is typical, excluded, or a matter of something else. Our critical robotics program builds on design and study examples, as well as contributions from the research community to articulate and exemplify concerns part of a critical robotics perspective (Ljungblad et al., 2018; Serholt et al., 2022). We now offer some of our own interdisciplinary work as exemplars, illustrating our own stand. For example, to complement the specific lab experiments that are common in robotics, we have conducted ethnographic studies addressing social and other challenges and overall experiences of using robotic products in homes and engaged in children's perspectives; a user group that is often missing in robotic research (Fernaeus et al., 2009; Gamboa, 2022; Ljungblad, 2023). We have studied existing professional drone pilot practices, and mapped out specific areas of research that are missing and need to complement current lab experiments and speculative scenarios of drone use in research (Ljungblad et al., 2021). We have studied how teaching with a social robot in a real school environment was experienced by students, raising awareness of the trouble that occurs in conversations with social robots and how this can affect the overall learning situation (Serholt et al., 2020). Our research has also looked into how a transportation robot used at a hospital can raise specific ethical issues and affect the experience of a work setting (Ljungblad et al., 2012; Nylander et al., 2012). We have also learned from artists and their practices how robotic artworks open for close encounters with the general public in the showroom, generating playfully ethical and aesthetic questions, beyond existing myths (Jacobsson et al., 2013). Finally, we have conducted critical and inclusive conceptual design work on drones intended for the showroom to open up a critical debate about the use of drones (Gamboa et al., 2023).

To develop this program, we take inspiration from Redstrom (2017). He describes how a research program can be characterized by both intent and unfolding, with a projection and a process that are intertwined. He builds his perspective on that of Imre Lakatos, viewing a research program as an overall framework, building on a set of series, and providing a foundation for future research within a specific worldview.

Redstrom (2017) emphasizes that theory in design is not a fixed or absolute entity but is continuously evolving and contested. It is shaped by ongoing debates, diverse perspectives, and new insights emerging from design practice and research. The overall goal of critical robotics is to strengthen the epistemological foundations of sound and desirable research that is excluded or exists on the margins, for example, due to not following a dominant research tradition. This we do by encouraging and nurturing additional reflective accounts and critical perspectives. So far, the program consists of a basic set of beliefs or constructs. We hope to see these evolving and growing in the field through design exemplars, case studies, and the development of mid-range theories. Our work, and its overarching context, represent some basic definitions connected to a critical robotics stand and methodology. In this chapter, we describe an initial set of constructs and invite other researchers to address these to strengthen design knowledge within HRI:

- Problematization
- Marginal perspectives
- Moving beyond user requirements
- The role of the designer

The above should be understood as interdisciplinary constructs to guide critical and ethical discourses in HRI research (Ljungblad et al., 2018; Serholt et al., 2022).

8.2 PROBLEMATIZATION

What is the essential research question for a specific robot project, and how does this question matter for the society? What is our role as researchers compared to industry and other knowledge-producing practices? Which stories of robots are we as researchers sharing and why (Brandão, 2021; Fernaeus et al., 2009)? Are robots approached as technological fixes to social problems (Sabanović, 2010)? Could specific design methods support us to reformulate the initial research question and avoid design fixation in research (Ljungblad, 2023)? When it comes to the more practical use of robots, the problematization concerns the initial research question and its context, the possible methods to question it, and setting a specific research direction. Problematization in critical robotics is about taking a stand to keep holistic and humanistic sensitivity and openness within the robotic research project. Specific methods (in design or research) may or may not support the researcher to view the situation from different angles and from the perspective of different stakeholders (Lupetti et al., 2021). Often, after gathering data, taking a critical look at the problematization and the initial questions will suggest the need to reframe the initial idea of a robotic solution or support (Ljungblad, 2023). Design processes of robotic art (Kac, 1997; Yolgormez & Thibodeau, 2022), critical design (Dunne & Raby, 2001; Pierce et al., 2015) and speculative design (Auger, 2014) are typically not oriented toward practical use, but concerned with raising questions and supporting ethical reflection. For such works, the problematization can be to question myths and assumptions of interactions with robots. Such directions may, for example, build on robotic social uselessness and the required human intervention of repair (Yolgormez & Thibodeau, 2022). It is also possible to introduce or apply theoretical perspectives and methods from other fields, building on specific research traditions. One such example is how to apply feminist theory and bring knowledge into pluralistic research practices in HRI (Winkle et al., 2023). This can support an alternative problematization and more norm-creative perspectives beyond a dominant research tradition.

The need for more situated knowledge perspectives in research was pointed out as early as 1988 by Donna Harraway (Harraway, 2013). Critical robotics research can look into how robotic products might change and disrupt socio-cultural life worlds (Hildebrand, 2022; Nørskov, 2022). As such, a researcher's ethical compass and belief system are important, to let nuanced sociological accounts of people's existing practices and their central values be interwoven into research (Dobrosovestnova et al., 2022). To gain novel and norm-creative perspectives, sometimes redesigning or even undesigning the imagined use of robots may be necessary. Such knowledge is also valid critical robotics research knowledge. It does not have to be about generating design implications for a robot, but can be framed as a springboard for a critical debate. For example, Lupetti and Van Mechelen (2022) worked with children in school with a deceptive robot to critically discuss and raise awareness of deceptive behaviors associated with societal myths and assumptions about robots. However, critical reflection on more practical use aspects of robots can also support a sound problematization.

8.3 MARGINAL AND NORM-CREATIVE PERSPECTIVES

The field of HRI, in its interdisciplinarity, has focused on finding ways of studying robots that can be generalized. The goal is to conform to the criteria of rigor commonly accepted in engineering or psychology research. However, design knowledge is typically less concerned with generalization, and has other criteria for what is considered rigorous. Marginal practices may reveal innovation opportunities, studying marginalized users may lead to more accessible, usable, and norm-creative solutions, and the best possible methods for design can be the methods uniquely created for the specific situation at hand. As pointed out by (Lupetti et al., 2021):

> In this process, design methods help bridge the gap between the technical research interests that drive most engineering approaches in robotic research with the actual sociocultural reality and needs of potential users that robots may interact with.
>
> (Lupetti et al., 2021)

Design methods can make use of marginal perspectives that are usually pushed to the margins by existing assumptions. This means to incorporate not only what is a central tendency, but to take a stand to look at the edges of what is considered accepted knowledge within a discipline (see e.g. Gemeinboeck & Saunders, 2023; Luria et al., 2020; Winkle et al., 2023). Design methods can make visible and make use of what is usually perceived as errors or undesirable effects in research. These may include breakdowns, failures, and collisions. It can be argued that exposing knowledge that is usually hidden is an essential process of research – we should not only focus on contributing with perspectives on what should be done, but also on what ought to be avoided. Design processes are complicated, naturally exploratory, and hardly ever based on a hypothesis. Rather, what is prioritized is a sense of the possibility to be found in deeper engagement with emerging knowledge – problematizing what may be generally taken for granted (Gaver et al., 2022).

The favorable view of emergence (the unexpected things that happen in research) can be celebrated in design, and supported by methods such as research *through* design. Within robotic research, this implies studying and designing robots in uncontrolled settings with occasionally uncomfortable assumptions. This allows for a variety of potentially unexpected interactions to surface. These can then be analyzed and discussed in detail not merely as hypothetical scenarios but as ill exceptions to the habitual expectation of the rule.

The acceptance of alternative methods and perspectives is of great importance to the inclusivity of robotic research. Marginal methods do not only give voice to unusual perspectives, they also have the potential to make the research more accessible to those who are not usually perceived or incorporated as researchers. This allows for the surfacing of alternative forms of knowledge that do not follow the standardized ways of conducting research, supporting norm-creative approaches where research goals, perspectives, and discussions can be formed by the experience of marginal and typically excluded groups (Ljungblad, 2023). As previously mentioned, (Nanavati et al., 2023) explicitly include a community-based participatory research method, making one of the community researchers into a co-author.

8.4 BEYOND USER REQUIREMENTS

Even if user-centered methods such as participatory design are used in a design process, this does not guarantee that the result will be desired or accepted by the intended users (Lee et al., 2016). In our special issue on Critical Robotics 2021 (Serholt et al., 2022), several key contributions pointed out the risks of neglecting complexities in human practices in favor of packaging user requirements as design implications. For example, if care practices are deconstructed to give form to well-defined technical problems, this can lead to mechanization of care (Maibaum et al., 2022) and exacerbate stereotypes of care workers (Dobrosovestnova et al., 2022). Similarly, ageism occurs when older adults are depicted as fragile, vulnerable, and burdensome care recipients in need of a robot (Burema, 2022). When focusing research on socio-emotional relationships with robots, researchers may miss the more intrinsic and humanistic aspects of reciprocity among humans (van Wynsberghe, 2022). Moreover, much robot research involves primary users rather than the different types of tertiary stakeholders. This holds true even if the latter will have a primary role in the potential use. Also often absent is the role of lay experts in the potential implementation of robots, and the impact of power-balancing stakeholders (Weiss & Spiel, 2022). Methods such as autoethnography of robotic products can go beyond user requirements to focus on lived experiences in family life (Gamboa, 2022). A robot artifact presented for a user may "steal the show" when the researcher may need to know more about people's existing practices, experiences, and everyday situations.

For example, in assisted feeding, this can lead to merely reactive responses to a robotic feeding device, instead of learning about what is important in the overall meal experience (Ljungblad et al., 2021). Similarly, a focus on merely functional requirements may lead to rejection if the robot clashes with social and aesthetic values (Ljungblad, 2023). Another related aspect to be considered in user requirement gathering is response bias, such as social desirability. This is where people try to please the researcher and answer something that is socially acceptable rather than give honest answers. The presence of a robot can sometimes distort what may in reality be considered socially acceptable or not.

There are also different types of goals within design activities that sometimes have less to do with user requirements. Concept-driven designs, such as the critical, speculative, and artistic lenses, may explore and focus on one specific aspect of interaction intended for a showroom (Koskinen et al., 2011). These may include breathing to question and expand upon possible modes of interaction from a pluralistic perspective (Gamboa et al., 2023). The different mindsets create different expectations of the user requirements as they may be intended to be experienced in different contexts, whether it is a showroom, or a personal conversation piece.

Finally, another aspect going beyond user requirements is to consider how methods are applied differently by different researchers (Boehner et al., 2007). Some data collection methods were developed as inspiration for design, rather than approaches for requirement gathering – for example, cultural probes. An additional risk comes with viewing ethnographic methods as mere tools to get at user requirements (and implications for the design of robots). Such perspectives do not do justice to the very rich insights that they can provide with regard to human practices (Dourish, 2006).

8.5 POSITIONALITY

> Practice should involve a disclosure of the researcher's position in the world, her or his goals, as well as the researcher's position in her or his intellectual and, to an appropriate extent, political beliefs.
>
> (S. Bardzell & Bardzell, 2011)

A reflexive stance is not necessarily one that seeks to remove bias, but rather to be mindful of the researcher's positionality. Researchers can also strengthen or even alter their own position, by having co-authors that view, steer, and reflect on the research process from a marginal, alternative, and norm-creative perspective, such as when people with disabilities are co-researchers and co-authors (see e.g. Fossati et al., 2023; Nanavati et al., 2023). In feminist studies, strong objectivity was coined by Harding (1995) as a way to delink a "neutrality ideal" from standards aiming to maximize objectivity. Her perspective draws on standpoint epistemologies, arguing that all research (also natural scientific) is shaped by politics, institutional structures, and the specific languages employed. Weak objectivism defends and legitimates the problematic institutions and practices ideals – including the idea that it is possible to be value-neutral, normal, natural, apolitical, and absent of gender coding. Strong objectivity, on the other hand, embraces the role of experience in producing knowledge. It advocates global and local social changes, pointing out the need for diversity in science to gain the value associated with multiple diverse perspectives. To build on Harding's (1995) ideas from a methodological perspective, we see the need for researchers to be open with their position and open up for pluralism. This also includes welcoming the employment of a multitude of methods, including from outside of the HRI field. Along with this, we argue the need to actively search for methods that can question stereotyped perspectives and biases. We believe that all researchers have a moral compass, and that we all can contribute when it comes to taking a stand. Researchers can learn much if they are clear on their own position and research aims. Furthermore, transparency of the research process and honest reflections on failures can support other researchers to know better how to apply methods. As designers, we can use methods that reveal failures early and learn from our own and other people's failures when learning and practicing methods.

8.6 INVITATION TO CONTRIBUTE

Doing critical robotics requires taking a reflective and occasionally uncomfortable look at the fundamental goals and approaches of our field. This is why we believe that creating and growing a platform for this type of research is essential. Overall, examples of such work include questioning one's own assumptions of design or practice (Baumer & Silberman, 2011); the discourse and writing traditions our fields adhere to (Pierce et al., 2015); or the institutions and the existing structures we are all part of and are influenced by every day (Winkle et al., 2023). This will typically also involve a need to clash with the dominant research paradigm (see e.g. Harding (1995)). It will mean engaging critically with specific political and institutional structures. It will also require greater acceptance with regard to what is considered rigorous scientific writing and methodological approaches. It requires a greater integration of marginal perspectives, along with a greater willingness to engage in norm-creative perspectives. This is the whole reason for this design program as we outline it here. Any design program needs exemplars of research that can be used for clarification and inspiration; the more we have, the better. This is important in order to make sustainable change in the field. We hope to encourage a wide variety of research activists to come together to support each other. We want to embrace the joy of discovery, failures, and fun and thought-provoking research. We want our work to engage with people and robots with humor, and in playful and critical ways. This is so the community of researchers and other stakeholders can build upon each other to become more than a discipline, turning into a growing and flourishing community. We hereby invite other researchers who align themselves with our stated values to contribute to the further exploration and development of critical robotics in HRI. We would like to see a discipline that engages constructively with failures and negative results so as to avoid the overly positive spin that publication bias puts on the myths of our field. The questions we pose, the methods we use, and the things we use, are all related to ethical and moral considerations (Verbeek, 2011), and there are many exciting ways to discuss and reflect on these.

Here are some suggestions of topics for future critical robotics research:

- Work that defends and explains criticality in robotics, as well as questions critical of robotics as a program (see e.g. (Csikszentmihalyi, 2022))
- Research that clarifies and extends our critical constructs of problematization, marginal perspectives, user requirements, and positionality
- Research that takes a stand against far-fetched and unrealistically deceptive robotic visions – for example, by pointing to communication issues in robotic research (see e.g. (Fernaeus et al., 2009))
- Novel design methods to incorporate and systematize feminist and other norm-creative perspectives
- Case studies and explorations where there are rich accounts of people and their lived experience
- Deconstructions of robotic products, i.e. presenting existing technological limitations and practicalities
- Norm-creative perspectives of robots, exploring different types of materializations, and very limited practical use of robots in favor of social or aesthetic use
- Design approaches to playfully identify design fixations, bias, and stereotyping aspects

We also envision contributions that go way beyond this – those that inspire, provoke, and critically reflect upon robotic research. We hope to see you out there!

ACKNOWLEDGMENTS

We would like to give special thanks to Sofia Serholt and Niamh Ni Bhroin for contributions, support, and encouragement in the early writing stages of this chapter. Thank you also to Michael Heron for reflections and language support, and Philippa Beckman for final proofreading.

REFERENCES

ACM code of conduct. (2023). ACM code of conduct. https://www.acm.org/code-of-ethics [Accessed: 2023, May 10].

Agre, P., & Agre, P. E. (1997). *Computation and human experience*. Cambridge University Press.

Auger, J. (2014). Living with robots: A speculative design approach. *Journal of Human-Robot Interaction*, *3*(1), 20–42.

Bardzell, J., & Bardzell, S. (2013). What is "critical" about critical design? In *Proceedings of the SIGCHI conference on human factors in computing systems* (pp. 3297–3306).

Bardzell, S., & Bardzell, J. (2011). Towards a feminist HCI methodology: Social science, feminism, and HCI. In *Conference on human factors in computing systems – Proceedings* (pp. 675–684). https://doi.org/10.1145/1978942

Baumer, E. P., & Silberman, M. S. (2011). When the implication is not to design (technology). In *Proceedings of the SIGCHI conference on human factors in computing systems* (pp. 2271–2274).

Bischof, A., Hornecker, E., Krummheuer, A. L., & Rehm, M. (2022, March). Re-configuring human-robot interaction. In *2022 17th ACM/IEEE International Conference on Human-Robot Interaction (HRI)* (pp. 1234–1236). IEEE.

Boehner, K., Vertesi, J., Sengers, P., & Dourish, P. (2007). How hci interprets the probes. In *Proceedings of the SIGCHI conference on Human factors in computing systems* (pp. 1077–1086).

Brandão, M. (2021). Normative roboticists: The visions and values of technical robotics papers. In *2021 30th IEEE international conference on robot & human interactive communication (RO-MAN)* (pp. 671–677).

Brando, M., Mansouri, M., & Magnusson, M. (2022). Responsible robotics. *Frontiers in Robotics and AI, 9*, 1–9.

Burema, D. (2022). A critical analysis of the representations of older adults in the field of human–robot interaction. *AI & Society, 37*(2), 455–465.

Csikszentmihalyi, C. (2022). An engineer's nightmare: 102 years of critical robotics. In *Presented at the "Re-Configuring Human-Robot Interaction", at the ACM/IEEE international conference on Human robot interaction.*

de La Bellacasa, M. P. (2011). Matters of care in technoscience: Assembling neglected things. *Social Studies of Science, 41*(1), 85–106.

Dignum, V. (2019). *Responsible artificial intelligence: How to develop and use AI in a responsible way*. Springer.

DiSalvo, C. (2015). *Adversarial design*. MIT Press.

Dobrosovestnova, A., Hannibal, G., & Reinboth, T. (2022). Service robots for affective labor: A sociology of labor perspective. *AI & Society, 37*(2), 487–499.

Dourish, P. (2006). Implications for design. In *Proceedings of the SIGCHI conference on Human Factors in computing systems* (pp. 541–550).

Dourish, P., Finlay, J., Sengers, P., & Wright, P. (2004). Reflective HCI: Towards a critical technical practice. In *CHI'04 extended abstracts on Human fac tors in computing systems* (pp. 1727–1728).

Dunne, A., & Raby, F. (2001). *Design noir: The secret life of electronic objects*. Springer Science & Business Media.

Fernaeus, Y., Jacobsson, M., Ljungblad, S., & Holmquist, L. E. (2009). Are we living in a robot cargo cult? In *Proceedings of the 4th ACM/IEEE international conference on human robot interaction* (pp. 279–280).

Fossati, M. R., Grioli, G., Catalano, M. G., & Bicchi, A. (2023). From robotics to prosthetics: What design and engineering can do better together. *ACM Transactions on Human-Robot Interaction, 12*(2), 1–24.

Friedman, B. (1997). *Human values and the design of computer technology*. Cambridge University Press.

Gamboa, M. (2022). Living with drones, robots, and young children: Informing research through design with autoethnography. In *Nordic Human-Computer Interaction Conference* (pp. 1–14).

Gamboa, M., Bayta, S. M. A., Hendriks, S., & Ljungblad, S. (2023). Wisp: Drones as companions for breathing. *Proceedings of the seventeenth international conference on tangible, embedded, and embodied interaction* (pp. 1–16).

Gaver, W., Krogh, P. G., Boucher, A., & Chatting, D. (2022). Emergence as a feature of practice-based design research. *Designing Interactive Systems Conference* (pp. 517–526). https://doi.org/10.1145/3532106.3533524

Gemeinboeck, P., & Saunders, R. (2023). Dancing with the nonhuman: A feminist, embodied, material inquiry into the making of human-robot relationships. In *Companion of the 2023 ACM/IEEE international conference on human-robot interaction* (pp. 51–59). https://doi.org/10.1145/3568294

Harding, S. (1995). "Strong objectivity": A response to the new objectivity question. *Synthese, 104*, 331–349.

Harraway, D. (2013). Situated knowledges: The science question in feminism and the privilege of partial perspective. In *Women, science, and technology* (pp. 455–472). Routledge.

Hildebrand, J. M. (2022). What is the message of the robot medium? Considering media ecology and mobilities in critical robotics research. *AI & Society*, *37*(2), 443–453.

Jacobsson, M., Fernaeus, Y., Cramer, H., & Ljungblad, S. (2013). Crafting against robotic fakelore: On the critical practice of artbot artists. In *Chi'13 extended abstracts on human factors in computing systems* (pp. 2019–2028).

Kac, E. (1997). Foundation and development of robotic art. *Art Journal*, *56*(3), 60–67.

Koskinen, I., Zimmerman, J., Binder, T., Redstrom, J., & Wensveen, S. (2011). *Design research through practice: From the lab, field, and showroom.* Elsevier.

Latour, B. (2008). What is the style of matters of concern. *Two lectures in empirical philosophy. Department of Philosophy of the University of Amsterdam, Amsterdam: Van Gorcum.*

Lee, H. R., Tan, H., & Sabanovic, S. (2016). That robot is not for me: Address ing stereotypes of aging in assistive robot design. In *2016 25th IEEE international symposium on robot and human interactive communication (RO-MAN)* (pp. 312–317). https://doi.org/10.1109/ROMAN.2016.7745148

Ljungblad, S. (2023). Applying "designerly framing" to understand assisted feeding as social aesthetic bodily experiences. *ACM Transactions on Human-Robot Interaction. 12*(2), 1–23.

Ljungblad, S., Kotrbova, J., Jacobsson, M., Cramer, H., & Niechwiadowicz, K. (2012). Hospital robot at work: Something alien or an intelligent colleague? In *Proceedings of the ACM 2012 conference on computer supported cooperative work* (pp. 177–186).

Ljungblad, S., Man, Y., Baytas, M., Gamboa, M., Obaid, M., & Fjeld, M. (2021). What matters in professional drone pilots' practice? An interview study to understand the complexity of their work and inform human-drone interaction research. In *Proceedings of the 2021 CHI conference on human factors in computing systems* (pp. 1–16).

Ljungblad, S., Serholt, S., Milosevic, T., Bhroin, N. N., Norgaard, R. T., Lindgren, P., Ess, C., Barendregt, W., & Obaid, M. (2018). Critical robotics: Exploring a new paradigm. In *Proceedings of the 10th Nordic conference on human-computer interaction* (pp. 972–975).

Lupetti, M. L., & Van Mechelen, M. (2022). Promoting children's critical thinking towards robotics through robot deception. In *HRI '22: Proceedings of the 2022 ACM/IEEE international conference on human robot interaction* (pp. 588–597).

Lupetti, M. L., Zaga, C., & Cila, N. (2021). Designerly ways of knowing in HRI: Broadening the scope of design-oriented hri through the concept of intermediate-level knowledge. In *Proceedings of the 2021 ACM/IEEE international conference on human-robot interaction* (pp. 389–398). https://doi.org/10.1145/3434073.3444668

Luria, M., Sheriff, O., Boo, M., Forlizzi, J., & Zoran, A. (2020). Destruction, catharsis, and emotional release in human-robot interaction. *Journal of Human-Robot Interaction*, *9*(4). https://doi.org/10.1145/3385007

Maibaum, A., Bischof, A., Hergesell, J., & Lipp, B. (2022). A critique of robotics in health care. *AI & Society*, *37*(2), 467–477.

Nanavati, A., Alves-Oliveira, P., Schrenk, T., Gordon, E. K., Cakmak, M., & Srinivasa, S. S. (2023). Design principles for robot-assisted feeding in social contexts. In *Proceedings of the 2023 ACM/IEEE international conference on human-robot interaction* (pp. 24–33).

Nørskov, M. (2022). Robotification & ethical cleansing. *AI & Society*, *37*(2), 425–441.

Nylander, S., Ljungblad, S., & Villareal, J. J. (2012). A complementing approach for identifying ethical issues in care robotics-grounding ethics in practical use. In *2012 IEEE RO-MAN: The 21st IEEE international symposium on robot and human interactive communication* (pp. 797–802).

Pierce, J. (2012). Undesigning technology: Considering the negation of design by design. In *Proceedings of the SIGCHI conference on human factors in computing systems* (pp. 957–966).

Pierce, J. (2014). Undesigning interaction. *Interactions*, *21*(4), 36–39.

Pierce, J., Sengers, P., Hirsch, T., Jenkins, T., Gaver, W., & DiSalvo, C. (2015). Expanding and refining design and criticality in HCI. In *Proceedings of the 33rd annual ACM conference on human factors in computing systems* (pp. 2083–2092).

Redstrom, J. (2017). *Making design theory*. MIT Press.

Richardson, K. (2015). *An anthropology of robots and AI: Annihilation anxiety and machines*. Routledge.

Sabanović, S. (2010). Robots in society, society in robots: Mutual shaping of society and technology as a framework for social robot design. *International Journal of Social Robotics*, *2*(4), 439–450.

Serholt, S., Ljungblad, S., & Ni Bhroin, N. (2022). Introduction: Special issue—Critical robotics research. *AI & Society*, *37*(2), 417–423.

Serholt, S., Pareto, L., Ekström, S., & Ljungblad, S. (2020). Trouble and repair in child–robot interaction: A study of complex interactions with a robot tutee in a primary school classroom. *Frontiers in Robotics and AI*, *7*, 46.

United Nations. (2023). The 17 goals. https://sdgs.un.org/goals [accessed 2023, May 10].

van Wynsberghe, A. (2022). Social robots and the risks to reciprocity. *AI & Society*, *37*(2), 479–485.

van Wynsberghe, A., & Sharkey, N. (2020). Special issue on responsible robotics: Introduction. *Ethics and Information Technology*, *22*, 281–282.

Verbeek, P. -P. (2011). *Moralizing technology: Understanding and designing the morality of things*. University of Chicago press.

Weiss, A., & Spiel, K. (2022). Robots beyond science fiction: Mutual learning in human–robot interaction on the way to participatory approaches. *AI & Society*, *37*(2), 501–515.

Winkle, K., McMillan, D., Arnelid, M., Harrison, K., Balaam, M., Johnson, E., & Leite, I. (2023). Feminist human-robot interaction: Disentangling power, principles and practice for better, more ethical HRI. In *Proceedings of the 2023 ACM/IEEE international conference on human-robot interaction* (pp. 72–82).

Yolgormez, C., & Thibodeau, J. (2022). Socially robotic: Making useless machines. *AI & Society*, *37*(2), 565–578.

9 Understanding Designerly Contributions

Nazli Cila
Delft University of Technology, Delft, Netherlands

The argument that "design is a discipline in its own right" marked a significant moment in the history of design research (Cross, 1982, 2018). Before this recognition, there was a strong push to develop design as a discipline primarily driven by cognitive scientists aiming to establish "a science of design," which employs empirical evidence and formal theorizing to model problem-solving (Ball & Christensen, 2019). However, in the 1980s, design began to assert its distinct ways of knowing, moving away from the logico-deductive approach to knowledge construction characteristic of positivistic Cartesian sciences, and differing from the humanities (Luck, 2019). Instead, design claimed its epistemics as a cognitive, social, and creatively reflective practice (Cooper, 2019).

Design plays a crucial role in Human–Robot Interaction (HRI) research and practice. Creative and critical design approaches are essential for envisioning (new) interactions with robots. While the HRI field is still developing its epistemological foundations and methodological frameworks, scholars increasingly recognize the need for devising new approaches and theoretical bases (Dautenhahn, 2018). Design, with its unique approach to generating knowledge, has been shaping and advancing the field. However, I argue that its full potential has yet to be realized in HRI.

Design knowledge encompasses both the formal, rooted in scientific and disciplinary principles, and the informal, grounded in intuitive and common-sense understandings of the world (Horvath, 2008). Its distinct modes of learning and understanding are commonly referred to as "designerly inquiry." Throughout this essay, I will advocate for the integration of designerly inquiry into the epistemology of HRI—that is, the understanding of what constitutes valid knowledge within a research domain and how such knowledge is acquired (Guba & Lincoln, 1994).

Yet, discerning what qualifies as a design contribution, what knowledge it yields, and its impact can prove challenging. At times, the field of design appears entangled in false dichotomies between research and practice, theory and application, academia and industry. These divides may also surface within designerly work in HRI. The Human–Computer Interaction (HCI) field has embarked on reconciling these divisions in fruitful ways, thereby enhancing the field's knowledge base and methodologies. In this essay, I aim to extend these efforts to HRI design, addressing these dichotomies to examine the role of design in HRI research, the possible outcomes of

DOI: 10.1201/9781003371021-9

This chapter has been made available under a CC-BY-NC-ND license.

designerly HRI work, and the contexts in which design unfolds and what we can learn from them. I intend to reflect on collaborative endeavors undertaken with my lovely colleagues, primarily Maria Luce Lupetti and Cristina Zaga, alongside other valued collaborators. Ultimately, as suggested by the title, I hope this essay serves as a starting point for understanding what constitutes "designerly" and serves as "a contribution" to HRI, thereby solidifying HRI as a distinct discipline with its own epistemological foundations and methodological frameworks.

9.1 FIRST DICHOTOMY: RESEARCH vs. PRACTICE

In the journey toward establishing design as a distinct discipline, a persistent perception has lingered: that makers exclusively make while theorists exclusively theorize, portraying these endeavors as entirely separate realms (Redstrom, 2017). Design research, with its twin pillars of "design" and "research," has often seemed at odds with itself, with design rooted in practical craftsmanship and industrial practice, and research situated in academic experimentation and reflection. Lloyd (2017) elucidated this apparent divide, highlighting the misguided assumption that research solely emphasizes objectivity, discourse, and analysis, while design is seen as solely concerned with shaping futures through suggestion, prototyping, and intervention across various levels of application.

In response to this perceived dichotomy, scholars have pursued different paths to reconcile the norms and values underlying scientific and design problem-solving processes. Some have sought to underscore the shared cognitive foundations of doing design and doing research (e.g., Farrell & Hooker, 2014), while others have endeavored to develop alternative theoretical frameworks to bridge the gap between design and research. One of the most influential endeavors in this regard was by Christopher Frayling. In his seminal address at the Royal College of Art in 1993, he delineated three ways in which research and design intersect within design research. The first of these, termed *Research for Design*, involves incorporating research activities—such as observation, measurement, interviews, literature review, analysis, and validation—into the design process, leveraging scientific and technological knowledge to inform design decisions. Notably, studies assessing the technical feasibility and usability of prototypes, as well as more recent participatory investigations into user requirements, have become integral components of doing Research for Design (Stappers and Giaccardi, 2017).

Recently, Maria Luce, Cristina, and I looked into the descriptions of the "design track" featured in the flagship ACM/IEEE International Conference on Human-Robot Interaction (from now on: HRI Conference) to have an understanding of the interpretation and evolution of design practices within the field (Lupetti, Zaga & Cila, 2021). Introduced in 2015, this track marked a significant turning point within the conference, representing a shift toward recognizing the importance of design-oriented approaches in HRI. During its inaugural year and subsequent edition, the design track showcased innovative developments in robot design, behavior, interaction paradigms, scenarios, and service designs. The focus was on creating and evaluating standalone robotic solutions, with insights gleaned from (user) research activities such as interviews, focus groups, and tests informing the design process and assessing the effectiveness of resulting robot designs. These endeavors in HRI represent valuable demonstrations of conducting Research for Design.

From 2017 onward, the description of the track underwent an expansion to include "research on the design process itself" and "critical reflection on the design process or methodology." This broadening signifies a significant step toward recognizing design as a discipline with its own distinct epistemology and methodology, demanding a dedicated inquiry. This evolution aligns with Frayling's second mode of research-design convergence: *Research into Design*. This mode encompasses research areas such as the history of design, aesthetics and design theory, and the analysis of design activity (Schneider, 2007). Similarly, Cross (2007) emphasizes the focus of this domain on elucidating the nature of design activity, design behavior, and design cognition. Much of his scholarly pursuit revolves around exploring "designerly ways of knowing," a concept extensively discussed within the literature on design research. This chapter of the book, in essence, can be seen as an exploration within this domain, seeking to unpack the implications of conducting design research for the field of HRI.

Frayling's last mode, arguably the most exciting for the current landscape of HRI, is *Research through Design* (RtD). RtD refers to a research approach which integrates methods and processes from design practice as legitimate forms of inquiry (Zimmerman et al., 2007). According to Stappers and Giaccardi (2017), at the core of RtD lies "the contribution of designerly activities and qualities to the *knowledge outcome*, especially those activities that introduce prototypes into the world, and reflect, measure, discuss, and analyze the effect, sometimes the coming-into-being, of these artifacts." Within this evolving paradigm, knowledge becomes inherent in the act of designing itself—acquired through active engagement in design activities and subsequent reflection—or is embodied within the design artifacts. As a result, the artifacts produced in this type of research can open unanticipated design spaces (Giaccardi et al., 2020), serve as vehicles for theory building (Koskinen et al., 2013), and provoke discussion around certain issues (Gaver, 2012). The insights yielded from the realization of such artifacts should not remain confined to the prototype in RtD, but ought to be integrated into disciplinary and cross-disciplinary platforms to enrich theoretical discourse (Stappers, 2007).

While RtD has claimed its validity and reliability as an approach to generate knowledge in the field of design (especially in HCI), its introduction into HRI is relatively recent. Critical reflections on methods, processes, and design outcomes are not common practice. However, notable efforts—such as from Luria, Zimmerman, and Forlizzi (2019), Lupetti (2017), Lee and Jung (2020), among others—have actively addressed the potential contributions of design to HRI and recognized the diverse ways in which design can enrich knowledge generation within the field.

Moving forward, for the HRI community to establish its epistemology of designerly HRI, it is necessary to embrace all these three modes of knowledge generation. There exists value in incorporating multiple paradigms of inquiry in research, each oriented toward distinct research programs and accommodating different objects and activities within its mode of inquiry (Kuutti & Bannon, 2014). This could involve, on a practical level, the inclusion of RtD and critical approaches to HRI in the HRI conference or other HRI venues. Moreover, on a fundamental level, it entails nurturing a culture of critical reflection on robotic design artifacts and processes to unlock the full potential of design as a powerful approach for the scientific growth of the discipline.

9.2 SECOND DICHOTOMY: THEORY vs. INSTANCE

In 2010, Erik Stolterman and Mikael Wiberg highlighted a prevalent trend at HCI conferences, where researchers presented artifacts showcasing new or improved forms of interactivity (Stolterman & Wiberg, 2010). These designs typically built upon previous work in the field, underwent some evaluation, and were presented alongside suggestions for future improvements. However, Stolterman and Wiberg argued that such presentations often failed to leave a lasting impact on fellow researchers because they did not intentionally address, challenge, or complement existing theoretical knowledge within interaction design research. Similarly, in 2017, Wolmet Barendregt and her colleagues identified a similar trend in the Child–Computer Interaction community, where many papers described the design process and evaluation of a specific artifact (Barendregt et al., 2017). These papers presented "design instances," which are highly situated, unique, and stand-alone products, among many other possible solutions. In line with Stolterman and Wiberg, the scholars argued that such papers made a smaller contribution to the field as it was challenging for other researchers to build upon the knowledge presented in them.

As design researchers primarily working at the intersection of design and HRI, Maria Luce, Cristina, and I were keenly aware of the same epistemological challenge within our domain. We observed that the role of robotic artifacts has predominantly been about addressing concrete problems, such as enhancing the user-friendliness of mobile robot mission specification systems or refining the social interactions of robots with humans. This observation aligns with the findings of Baxter and his colleagues, who demonstrated that 96 out of 101 papers from the HRI conference between 2013 and 2015 aimed to provide an exploration around a human-centered phenomenon or improve the functionalities and application of a robot (Baxter et al., 2016). This is not to undermine the value of such work; these artifacts have undeniably offered technologically advanced and valuable solutions to specific problems. However, we were mainly concerned about the potential risk that only a fraction of the knowledge gained from addressing one problem could be transferred to the next one.

Drawing inspiration from the study of Barendregt et al. (2017), we continued our examination of the HRI Conference and investigated the extent to which HRI Conference papers centered around artifacts (Cila, Zaga & Lupetti, 2021). We reviewed 587 full papers spanning the years 2006 to 2020. Our analysis revealed that the distribution ranged from 30% to 54%, with an average of 41.2% across all years, indicating a substantial proportion of HRI Conference papers focused on specific robotic artifacts. Furthermore, we sought to assess the impact of these artifact-centered papers by comparing their citation numbers with other types of papers. Our findings indicated that artifact-centered papers indeed tended to have a lower impact within the field.

In design theory, research has demonstrated that the development of theoretical frameworks predicts significant increases in citation counts (Beck & Chiapello, 2018). This phenomenon is logical since theories are formulated at a high-level of abstraction, allowing for their application across various contexts, making them generalizable. In the field of HRI, theories play a crucial role in establishing a foundational understanding of HRI, elucidating how specific design decisions influence these interactions, and exploring novel mechanisms or computational tools to enhance them (Jung & Hinds, 2018). However, given the applied nature of the field and the multitude of engineering and design challenges to address, it is unreasonable to expect all HRI research to operate at the level of theories. Design instances are needed to solve specific HRI problems.

In recent years, the HCI field has recognized that valid and reliable knowledge can be elicited from design instances, contributing meaningfully to the field. This "intermediate-level knowledge" is a representation of knowledge in-between general theories and design instances (Höök & Löwgren, 2012). It provides explanations beyond individual artifacts but does not aim to formulate generalizable theories. More abstract than specific instances yet less general than theories, intermediate-level knowledge serves as a valuable tool for fostering discussions on paradigmatic systems and exemplary artifacts that transcend specific functions and aesthetics (Stolterman & Wiberg, 2010). This enriches the discourse on interaction design epistemology and enhances the rigor of design-oriented research. Examples of intermediate-level knowledge include patterns, guidelines, annotated portfolios, methods and tools, experiential qualities, design heuristics, criticism, strong concepts, manifestos, design programs, and bridging concepts (Höök & Löwgren, 2012).

The adoption of these diverse knowledge forms in HRI is still in its early stages. In 2021, we conducted an extensive overview of their utilization in HRI (Lupetti et al., 2021). A few examples include the creation of "design patterns" to describe robot–child interactions in specific contexts (Kahn et al., 2008), the application of "heuristics" such as continuous actions and boundary signaling to enhance robots' social acceptability (Dautenhahn, Ogden & Quick, 2002), and "criticism" concerning the representation of the divine in robots (Trovato et al., 2018). Additionally, we also elicited "strong concepts" from prominent artifact-centered papers in the field (Cila et al., 2021) and explored the use of "annotated portfolios" as a means to elucidate the values, concepts, and assumptions typically implicit in existing robotic artifacts, which are often overlooked as knowledge contributions (Lupetti et al., 2022).

Embracing intermediate-level knowledge forms, alongside theory and instance development, is crucial for the advancement of disciplines in which design plays a central role. While certain intermediate-level knowledge forms such as methods and guidelines are already familiar within the HRI community, others like criticism and annotated portfolios are gaining traction. We view the latter as particularly important for grasping the conceptual significance of robotic artifacts developed within HRI research. Specifically, we believe that conceptual investigations of HRI artifacts, focusing on their embodiments and behaviors, can challenge preconceived notions of how robots should appear, behave, and exist, thereby shaping the envisioning of our future alongside robots (Lupetti et al., 2021).

9.3 THIRD DICHOTOMY: ACADEMIA vs. INDUSTRY

The final dichotomy to address in understanding designerly contributions lies in the differences between practicing HRI in academia and in industry. In the realm of the latter, interaction designers and UX designers anticipate and explore user needs, behaviors, and cognitive abilities, leveraging these insights to craft robotic artifacts that prioritize utility, usability, and user-friendliness. Their design considerations span the robots' operating system, interface, form, configuration, sound, and movement. In academia, HRI researchers study the factors that distinguish successful HRI, constructing frameworks that delineate key interaction qualities.

However, much of HRI research to date has been confined to controlled laboratory environments, often involving a single human interacting with a single robot (Jung & Hinds, 2018), lacking connection with real user populations and environments (Matarić, 2018). This detachment from real-world complexities, such as the need for attention to detail, reliability, and robustness in real-world deployments, presents challenges in integrating HRI research findings into robot design practice. Conversely, in the industry realm, these concerns take precedence, often under strict confidentiality requirements imposed by companies or specific contexts, such as military applications. This confidentiality may hinder the sharing of internal processes, findings, and procedures with a broader audience. Moreover, the unpredictable nature of real-world environments introduces complexities and uncertainties that may contradict the structured and robust approaches required for developing theoretical frameworks.

These challenges highlight the bidirectional flow of knowledge between academia and practice in HRI design work, with barriers existing both in the translation of research insights into robot development and in the integration of real-world experiences into theoretical frameworks. This issue, commonly termed the "gap problem," is not unique to HRI but is also prevalent within the HCI community. Studies have repeatedly demonstrated that while practitioners acknowledge the value of theoretical insights from research, they are seldom incorporated in design practice (Colusso et al., 2017; Goodman, Stolterman & Wakkary, 2011; Gray, Stolterman & Siegel, 2014). Fallman and Stolterman (2010) attribute this gap to three factors: relevance (addressing problems and themes that are important to professionals), applicability (being able to utilize results in the form of new knowledge and methods), and accessibility of research findings (presenting research in an understandable way).

Recently, I collaborated with Lely, a global leader in the development of dairy farming robots, with the aim of improving the communication behaviors of their robots with farmers (Cila et al., 2024). Lely possesses extensive expertise in their technology, robots, and the specific farm tasks (e.g., milking, feeding, or cleaning the cows), and is intended to broaden their knowledge on HRI by integrating more academic research into their processes to design better farmer–robot interactions. However, they reported encountering challenges in translating the high-level abstract knowledge typically produced by academic HRI research into actionable insights for their specific context.

Similar to the previous section, there seemed to be a gap between the abstract knowledge generated by HRI research and the practical, context-specific interaction design knowledge utilized by Lely. To bridge this gap, we once again turned to intermediate-level knowledge. Specifically, we developed "design guidelines" aimed at enriching Lely's specialized understanding of farming robots and the farm environment with relevant academic insights. These guidelines represented a synthesis of the specific needs of the dairy farming context, the company's values, and pertinent theoretical knowledge concerning best practices in HRI.

During this process, we uncovered opportunities for a reciprocal exchange between HRI theories and robot development practice. For example, insights from HRI theory elucidated the sociotechnical complexity of the robotized dairy farming context, pinpointing points of attention, and served as a tool for crafting interview questions for farmers. In the resultant design guidelines, the theory also manifested itself in concrete takeaways and best practices. Conversely, the knowledge from the dairy farming context guided the search for relevant HRI theories and validated their applicability to this specific setting. It also informed user research, indicating which issues were crucial to discuss during interviews with stakeholders. All information and examples incorporated in the guidelines were rooted in real-world phenomena and experiences, a quality that was a significant asset during the evaluation process within the company. Similarly, insights from Lely's professional practice ensured the end design outcome seamlessly aligned with the company's workflow and culture.

In this project, our objective was to narrow the gap between HRI theory produced in academia and HRI practice followed in industry. Achieving this required close collaboration with the company within a particular real-world context. By doing so, we addressed a problem of significance to a specific robot development practice, devised a format familiar to the company and actionable for communicating research insights, and presented these insights in an understandable manner. These efforts enabled us to navigate around the factors identified by Fallman and Stolterman (2010) that contribute to the divide between academia and industry. Furthermore, we aimed to underscore the dynamic interplay between designerly HRI research and practice, highlighting the various ways in which they can complement each other. This project exemplified the dialectical nature of designerly HRI work, where theory and practice interact reciprocally, thereby enriching HRI epistemology.

9.4 CONCLUSION

In essence, understanding designerly contributions to the field of HRI requires grappling with complex dichotomies and forging pathways for reconciling these. Through the lens of research vs. practice, theory vs. instance, and academia vs. industry, we uncover the intricate interplay between knowledge generation, application, and dissemination. While design has emerged as a distinct discipline with its own epistemological foundations, its integration into HRI remains a work in progress. By wholeheartedly embracing the potential that design holds, we possess the power to mold the future of HRI, infusing them with creativity, insight, and impact. This call to action extends as an open invitation to the HRI community—a collective endeavor to embrace design to expand and enrich HRI epistemology, while shaping a future where HRI flourish with newfound richness and depth.

REFERENCES

Ball, L. J., & Christensen, B. T. (2019). Advancing an understanding of design cognition and design metacognition: Progress and prospects. *Design Studies*, *65*, 35–59.

Barendregt, W., Torgersson, O., Eriksson, E., & Börjesson, P. (2017, June). Intermediate-level knowledge in child-computer interaction: A call for action. In *Proceedings of the 2017 conference on interaction design and children* (pp. 7–16).

Baxter, P., Kennedy, J., Senft, E., Lemaignan, S., & Belpaeme, T. (2016, March). From characterising three years of HRI to methodology and reporting recommendations. In *2016 11th ACM/IEEE international conference on human-robot interaction (HRI)* (pp. 391–398). IEEE.

Beck, J., & Chiapello, L. (2018). Schön's intellectual legacy: A citation analysis of DRS publications (2010–2016). *Design Studies*, *56*, 205–224.

Cila, N., González González, I., Jacobs, J., & Rozendaal, M. (2024, March). Bridging HRI theory and practice: Design guidelines for robot communication in dairy farming. In *Proceedings of the 2024 ACM/IEEE international conference on human-robot interaction* (pp. 137–146).

Cila, N., Zaga, C., & Lupetti, M. L. (2021, June). Learning from robotic artefacts: A quest for strong concepts in human-robot interaction. In *Proceedings of the 2021 ACM designing interactive systems conference* (pp. 1356–1365).

Colusso, L., Bennett, C. L., Hsieh, G., & Munson, S. A. (2017, June). Translational resources. In *Proceedings of the 2017 conference on designing interactive systems*. ACM.

Cooper, R. (2019). Design research–Its 50-year transformation. *Design Studies*, *65*, 6–17.

Cross, N. (1982). Designerly ways of knowing. *Design Studies*, *3*(4), 221–227.

Cross, N. (2007). From a design science to a design discipline: Understanding designerly ways of knowing and thinking. In *Design research now* (pp. 41–54). Springer.

Cross, N. (2018). Developing design as a discipline. *Journal of Engineering Design*, *29*(12), 691–708.

Dautenhahn, K. (2018). Some brief thoughts on the past and future of human-robot interaction. *ACM Transactions on Human-Robot Interaction (THRI)*, *7*(1), 1–3.

Dautenhahn, K., Ogden, B., & Quick, T. (2002). From embodied to socially embedded agents–implications for interaction-aware robots. *Cognitive Systems Research*, *3*(3), 397–428.

Fallman, D., & Stolterman, E. (2010). Establishing criteria of rigour and relevance in interaction design research. *Digital Creativity*, *21*(4), 265–272.

Farrell, R., & Hooker, C. (2014). Values and norms between design and science. *Design Issues*, *30*(3), 29–38.

Gaver, W. (2012, May). What should we expect from research through design? In *Proceedings of the SIGCHI conference on human factors in computing systems* (pp. 937–946).

Giaccardi, E., Speed, C., Cila, N., & Caldwell, M. L. (2020). Things as co-ethnographers: Implications of a thing perspective for design and anthropology. In *Design anthropological futures* (pp. 235–248). Taylor & Francis.

Stappers, P. J., & Giaccardi, E. (2017). Research through design. In *The encyclopedia of human-computer interaction* (pp. 1–94). The Interaction Design Foundation.

Goodman, E., Stolterman, E., & Wakkary, R. (2011, May). Understanding interaction design practices. In *Proceedings of the SIGCHI conference on human factors in computing systems* (pp. 1061–1070).

Gray, C. M., Stolterman, E., & Siegel, M. A. (2014, June). Reprioritizing the relationship between HCI research and practice: Bubble-up and trickle-down effects. In *Proceedings of the 2014 conference on designing interactive systems* (pp. 725–734).

Guba, E. G., & Lincoln, Y. S. (1994). Competing paradigms in qualitative research. In *Handbook of qualitative research* (pp. 105–117). Sage Publications, Inc.

Höök, K., & Löwgren, J. (2012). Strong concepts: Intermediate-level knowledge in interaction design research. *ACM Transactions on Computer-Human Interaction (TOCHI)*, *19*(3), 1–18.

Horvath, I. (2008). Differences between 'research in design context' and 'design inclusive research' in the domain of industrial design engineering. *Journal of Design Research*, *7*(1), 61–83.

Jung, M., & Hinds, P. (2018). Robots in the wild: A time for more robust theories of human-robot interaction. *ACM Transactions on Human-Robot Interaction (THRI)*, *7*(1), 1–5.

Kahn, P. H., Freier, N. G., Kanda, T., Ishiguro, H., Ruckert, J. H., Severson, R. L., & Kane, S. K. (2008, March). Design patterns for sociality in human-robot interaction. In *Proceedings of the 3rd ACM/IEEE international conference on Human robot interaction* (pp. 97–104).

Koskinen, I., Zimmerman, J., Binder, T., Redstrom, J., & Wensveen, S. (2013). Design research through practice: From the lab, field, and showroom. *IEEE Transactions on Professional Communication*, *56*(3), 262–263.

Kuutti, K., & Bannon, L. J. (2014, April). The turn to practice in HCI: towards a research agenda. In *Proceedings of the SIGCHI conference on human factors in computing systems* (pp. 3543–3552).

Lee, W. Y., & Jung, M. (2020, March). Ludic-hri: Designing playful experiences with robots. In *Companion of the 2020 ACM/IEEE international conference on human-robot interaction* (pp. 582–584).

Lloyd, P. (2017). From design methods to future-focused thinking: 50 years of design research. *Design Studies*, *48*, A1–A8.

Luck, R. (2019). Design research, architectural research, architectural design research: An argument on disciplinarity and identity. *Design Studies*, *65*, 152–166.

Lupetti, M. L. (2017). Shybo–design of a research artefact for human-robot interaction studies. *Journal of Science and Technology of the Arts*, *9*(1), 57–69.

Lupetti, M. L., Zaga, C., & Cila, N. (2021, March). Designerly ways of knowing in HRI: Broadening the scope of design-oriented HRI through the concept of intermediate-level knowledge. In *Proceedings of the 2021 ACM/IEEE international conference on human-robot interaction* (pp. 389–398).

Lupetti, M. L., Zaga, C., Cila, N., Luria, M., Hoggenmüller, M., & Jung, M. F. (2022, March). 2nd international workshop on designerly HRI knowledge. Reflecting on HRI practices through annotated portfolios of robotic artefacts. In *2022 17th ACM/IEEE international conference on human-robot interaction (HRI)* (pp. 1269–1271). IEEE.

Luria, M., Zimmerman, J., & Forlizzi, J. (2019). Championing research through design in HRI. arXiv preprint arXiv:1908.07572.

Matarić, M. (2018). On relevance: Balancing theory and practice in HRI. *ACM Transactions on Human-Robot Interaction (THRI)*, *7*(1), 1–2.

Redstrom, J. (2017). *Making design theory*. MIT Press.

Schneider, B. (2007). Design as practice, science and research. In *Design research now* (pp. 207–218). Birkhäuser: Basel.

Stappers, P. J. (2007). Doing design as a part of doing research. In *Design research now* (pp. 81–91). Birkhäuser: Basel.

Stolterman, E., & Wiberg, M. (2010). Concept-driven interaction design research. *Human–Computer Interaction*, *25*(2), 95–118.

Trovato, G., Lucho, C., Huerta-Mercado, A., & Cuellar, F. (2018, March). Design strategies for representing the divine in robots. In *Companion of the 2018 ACM/IEEE international conference on human-robot interaction* (pp. 29–35).

Zimmerman, J., Forlizzi, J., & Evenson, S. (2007, April). Research through design as a method for interaction design research in HCI. In *Proceedings of the SIGCHI conference on human factors in computing systems* (pp. 493–502).

10 Toward a Future Beyond Disciplinary Divides

Cristina Zaga
University of Twente, Enschede, Netherlands

Have you ever read an academic paper about robots? There is a high chance that the introduction features a sentence like this: Robots are increasingly being developed and will be integrated into the societal fabric in the next few years.

Such formulaic rhetoric might sound like delusional techno-positivism, especially when much of the robot the human–robot interaction (HRI) community envisioned in the last 20 years failed to become a consumer product (Hoffman, 2019). Similarly, if we look at the media discourse around robots, the techno-positivist questions and dystopic fears regarding when robots will come and how they remain: Are robots going to take over our jobs (Drenik, 2022; Orduña, 2021)? Are they going to be moral beings (Savage, 2019)? Are they going to upgrade our failed humanity? (Richardson, 2024)

The rhetoric of automation and marvel, which we have discussed at length in the book, is alive and kicking inside and outside academia. It is fueled by a media frenzy that does not cease, particularly after the semi-commercial introduction of large language models (LLMs) (Toso et al., 2023). The need for community reflection on what a robot should be and its societal implications is higher than ever (Lupetti et al., 2021).

Our book is a critical compass beyond the current robot's hype and frenzy, and it positions itself as a hands-on manual for developing technology for a future worth wanting.

The book's ten chapters show how HRI designs desirable human–robot relations rather than inevitable robot-dystopias (Vallor, 2016). By providing a palette of best practices, positions, methods, tools, and techniques from academia and industry, our book immerses the reader in a broader plurality of epistemologies and methodologies, slowly broadening the field and opening new avenues of meaningful technology production.

Hence, the book showcases works that could be seen as a sort of résistance (the resistance), critically shaping robotic technology otherwise, steering research toward doing the "right thing" as opposed to doing "things right" (Luria et al., 2020).

This chapter thus focuses on what we can learn from the book and future outlooks for the HRI field.

First, we reflect on what the reader (academic or non-academic) can learn from the book. Second, following our wish to broaden and loosen disciplinary boundaries, we discuss what we can learn from HRI and how the field may move forward, proposing a set of ontological, political, and epistemological reframes. Finally, we conclude with our wishes for the future(s) of meaningful, just, and rewarding human–robot relations.

DOI: 10.1201/9781003371021-10
This chapter has been made available under a CC-BY-NC-ND license.

WHAT CAN WE LEARN FROM THE BOOK? REFLEXIVITY, QUESTIONING, AND CURIOSITY

With more than thirty methods divided into four thematic sections, a reader might learn about (and hopefully engage with) many ways of designing robots and human–robot relations. However, the book is not a collection of recipes to try out. This volume offers a window into designerly ways of knowing, doing, and making, which have transformative consequences on how we practice research, engage with materiality, and make sense of human and non-human agency.

As such, a reader, academic or non-academic, is primarily compelled to explore ways to engage as a ***reflective practitioner*** (i.e., a researcher that embraces iterative reflection about action and role in research and iterative reflection through the generation of artifacts and engagement with materiality) (Schön 2017, Schön & DeSanctis, 1986). Shifting toward **reflexive** practice well combines with the open-ended nature of design, which follows abductive reasoning (Dorst, 2011) and focuses on situatedness (Gero & Milovanovic, 2023),

Engagement with design is not a dead disciplinary end, though. Design is often referred to as a binding glue among other disciplines (Ozkaramanli et al., 2022), like philosophy and science and technology studies, which may support the sense-making of the socio-technical facets of HRI and the inherently political act of designing. Therefore, one of the book's main takeaways is **questioning** the status quo of HRI from various perspectives. Chapters 2 and 7 spur us to go beyond current ways to intend appearance, expressivity, and materiality convenience to embrace more nuanced, aesthetic, and culturally grounded robot designs. While the entirety of Chapters 6 and 9 questions the politics of HRI, its narrative, and imaginaries invite the readers to question deep-seated narratives we overtly or covertly about robots. As we learn from Chapter 4, being a reflective practitioner means considering the mutual shaping between robots and society during the design process, refraining from considering societal implications as an afterthought.

The diversity of approaches, methods, and tools could be exciting or overwhelming, depending on how familiar a reader is with design, the humanities, and other generative disciplines. Particularly for engineering and computer scientists, the holistic, fuzzy, and abductive nature of design could be disorienting and difficult to reconcile with more empirically oriented practices. At the same time, the methodological palette provided in the book is fuel for **curiosity**. In particular, Chapter 2 focuses on approaches such as quotidianity, body-storming, and ready-made prototyping, tapping on the familiar and unfamiliar simultaneously is an excellent port of entry to engage with materiality, making and conceptualizing no matter what a reader's background is. However, most methods featured in the books are hybrids between disciplinary worlds, byproducts of openness, and mutual learning between practices.

Overall, the book opens a window to designerly ways of knowing, socio-technical reflections, and methodological eclectism, which contribute to developing reflexivity about practicing HRI, questioning the current HRI status quo, and a generalized curiosity about other epistemologies and methodologies.

LEARNING FROM AND REFRAMING HRI

What can other fields we learn from HRI? What can HRI learn from going beyond disciplinary silos? As shown in all the chapters of this book, HRI is arguably the first field to fully understand what it means to interact with and design for technology perceived and interacted with as animate, agentic, and social. HRI provides crucial intermediate-level knowledge about designing robots, even though the field is only sometimes aware of it (Lupetti et al., 2021) or attentive to documenting it (Cila et al., 2021).

A good example is the work led by Nazli Cila on strong concepts in HRI, which are generative design elements of principles that guide design within and beyond the field. By analyzing papers focusing on HRI design, we have identified two strong concepts: grounding, a fundamental communicative feature that aligns communication between agents through various ways of expressing gaze, and *same vs. others*, an ontological feature that describes the spectrum of design from human likeness to abstract non-anthropomorphism. Through this work, we have identified patterns and ways of designing in HRI holistically, which others can use in and outside the field to generate design. At the same time, engaging in such exercises to make sense of the knowledge generated in HRI through the lens of intermediate-level expertise can help us define what a robot is and how a robot should be and employ these design approaches to other design challenges even outside the realm of HRI, for example, in HCI when designing chatbots and LLMs-based agents.

If HRI has such a rich knowledge production with a high potential for informing other disciplines, why must we reframe the field?

Our work (Cila et al., 2021; Lupetti et al., 2021) shows that HRI still needs help making sense of the knowledge generated from the process and making robotic artifacts. Practices such as Research Through Design (RtD, see Chapter 8) are still uncharted territory, and it is challenging to organize the HRI design knowledge so that there are clear links between artifacts and epistemology.

Moreover, HRI produces technological artifacts that create new practices, social habits, and ways of living, impacting people's identity, relations, and societal position. Therefore, we must rethink how we practice HRI to put a societal impact at the core of our focus.

To this end, the field has seen the rise of critical approaches that ask to reconsider the premises of HRI itself. Many researchers and designers are called to carefully examine the socio-technical frictions implicitly embedded in a robot's design and try to counteract the dehumanizing and marginalizing effects of HRI through research and design. For example, Winkle et al. (2021) explored how to design robots to challenge current gender-based norms of expected behavior, suggesting that intervening in the design of roles, behavior, and embodiment could reduce gender biases. Tanqueray et al. (2022) ventured into gender-sensitive design approaches inspired by human-centered design and governance design to develop socially assistive robots for peripartum depression screenings. Hou et al. (2024) show how power and power dynamics are ingrained in human–robot relations and how subtle and often undocumented they are, calling for collaborative outlining ways to consider power in HRI.

Participatory approaches have been increasingly explored to practice a more inclusive HRI (Weiss & Spiel, 2022), but without tackling the inherent inclusivity risks of participatory design, i.e., not acknowledging power dynamics, costs of participation, and paternalistic tendencies (Zaga & Lupetti, 2022) and without a coherent way to describe the approaches like in other fields (i.e., child–computer interaction) (Walsh et al., 2013).

Notably, HRI scholars have shown that HRI tends to study the effects of robots and involve people in the process, with generalizable and interchangeable actors paying little attention to diversity and inclusion (Lee et al., 2022). Perugia and Lisy (2023) illustrate this phenomenon, highlighting how the HRI literature fails to account for non-binary, transgender, gender non-conforming, and gender fluid participants in studies exploring the impact of robot genderness.

Moreover, few are the participatory studies and practices that support generative and critical reflection of the HRI paradigms and phenomena, with the remarkable exception of a few instances, such as Lupetti and Van Mechelen work to help children's reflection about deceptions in HRI (Lupetti & Van Mechelen, 2022), the work of Luria et al., using performances to discuss robot futures critically (Luria et al., 2020) and Lee et al., work reframing the status quo of robots for dementia with the communities (Lee et al., 2023).

Related works show that the HRI community needs to engage with research paradigms and methods that tackle social justice. Most studies focus on understanding the effects of HRI on justice through empirical research. Still, it must provide designer knowledge informing design methods, guidelines, or patterns to guide inclusive and social justice-oriented practices. We can learn a lot from HRI, and HRI is learning a lot from other disciplines.

HRI *is* eagerly spilling out its disciplinary silos (and the ivory tower of academia as well) to learn from and share different and more inclusive ways of producing knowledge. HRI is grappling with what a robot is, what a robot should be, and how to make sure that robots are not yet another means of oppression negatively impacting our environment.

Yet, moving forward, the field needs to interrogate the foundation of its ontology (what exists, what does not, and how we make sense of them), its epistemology (its theory and practice of knowledge production), and its politics to realize its commitments fully.

THREE REFRAMES TOWARD A FUTURE BEYOND DISCIPLINARY DIVIDES

To fully realize the evolutionary broadening we seek and, ultimately, the societal impact we wish for, we must collectively confront our frame, i.e., our interpretative schemata, the way we look, make sense of our work, and use the compass we use to produce knowledge through research and design. Reframing standard practices in design processes (Kolko, 2010) and reformulating our schemata through collective sense-making of our research endeavors is necessary. Below, we offer my take on reframing HRI, focusing on ontological, political, and epistemological reframing.

Ontological Reframe: Connecting Relationality with Pluriversality

As we have seen in Chapters 4–6. HRI rests on a way of designing that is tied to the industrial era, developing what is "right" to solve "problems." Connecting to the essay of Sara Ljungblad and Mafalda Gamboa (Chapter 9) and as many scholars have highlighted (Frauenberger, 2019; Giaccardi & Redström, 2020; Wiltse, 2020), robots are a new type of artifact that mediate human experience through alterity relations (Latour & Weibel, 2005; Verbeek, 2005) and are entities in assemblages that constitute each other. From this perspective, it is ready to challenge the primary mode of making technology and robots: human-centered design. Even though one might argue that human-centered design just got traction in HRI (Lee et al., 2022), the nature of human–robot relations challenges this paradigm and approach and calls for alternatives.

It follows that the socio-technical and systemic nature of HRI relations calls to go beyond the binary distinction heritage of Cartesiam dualism (i.e., understanding the domain of reality as opposed to entities, like mind and body (Hawthorne, 2007)) between what is human and non-human and to adopt a more holistic perspective beyond the anthropocentric one. Human-centered design is focused on understanding humans as "distinct and individual subjects" (Forlano, 2017) from the perspective of the lived experience of humans. It is thus ill-equipped to fully consider the non-human. As a reaction to these challenges, designers are increasingly adopting relationality, which stands for the interconnected and co-shaping of all things, as well as so-called beyond-human-centered, more-than-human (Wakkary, 2021) and post-human design methods (Forlano, 2017), which focus on approaches that acknowledge and critically extend agency to non-human actors, including nature, animals, and artificial things (Cila et al., 2017).

HRI has pioneered practices that extend reflections about agency to non-humans (robots). Still, we have yet to understand theories outside HRI and their ontological implications fully.

As shown in Chapters 6 and 9, what HRI needs to do going forward, we argue, is to integrate relationality and confront HRI scholarship with the post-human/more-than-human discourse. On this, a partnership between HCI, which has inaugurated a new wave of its scholarship centered on relationality (Frauenberger, 2019), and design research would help navigate the ontological reframe.

Further, there is a lot to learn about relationality from the methods in the book, yet perspectives and criticisms from the scholarship from the Global South (Escobar, 2018) are still missing. Thus, we wonder: What about the humans who are not considered as such? How do we integrate the dehumanized perspectives and set them at the margins? How do we deal with Western values and categories often used to describe and understand people constructed as "others?"

Therefore, we argue that HRI—as it has already started doing—should explore how to integrate the so-called pluriversal perspective (Escobar, 2018) that connects relationality with efforts to counteract colonialism (Van Amstel, 2023), marginalization, and dehumanization to engage in practices of radical participation and communality for socio-cultural change (Noel et al., 2023). As many authors advocate (Arista et al., 2021; Birhane & van Dijk, 2020; Lewis et al., 2020), shifting our ontological perspective implies seeing robots as *mediating agents* embedded in a relationship with humans, animals, and nature.

Political Reframe: Toward Justice, Belonging, and Care

The last five years have seen a surge of scholarship highlighting the impact of HRI's politics, particularly regarding gender, racial, and disability matters, on design and society.

Research in HRI shows that the embodiment of a robot (i.e., how a robot looks), the behavior (verbal or nonverbal), and the role of a robot, HRI, can be perceived as gendered or racialized, perpetuating stereotypes and dehumanization that we observe between humans. HRI designers use anthropomorphism techniques or leverage the tendency of people to anthropomorphize robots to establish and sustain interaction between people and robots (Perugia et al., 2022). However, these design choices impact human–robot relations regarding gender and race.

Eyssel and Hegel (2012) demonstrate that gender stereotypes also apply in HRI. In her recent work, Perugia et al. (2022) further detail the granularity of the shape of people's bias to respect the robot's embodiment features. Reflection about gendering robots connects with HCI research that shows how smart assistant voices modeled onto female persona support acceptability and trust by reinforcing homogeneity in language and culture (Lee et al., 2021).

Bartneck et al. (2018) have shown that people automatically identify robots as racialized entities and apply racial biases. Strait et al. (2018) have shown that people more frequently dehumanize humanoid robots racialized as Asian and Black than they do of robots racialized as White. Plus, people seem to degrade humanoid robots with a gynoid form.

The robots' role and behaviors for specific social interventions (e.g., care) may also reinforce discriminatory and outdated models, nudging to "normative" human behaviors and policing what it means to be human. For example, Spiel et al. (2019) criticize how social robots for autistic children "*embody normative expectations of a neurotypical society, which predominantly views autism as a medical deficit in need of 'correction*" (page 38, abstract)

This research has sparked a renewed engagement with scholarship and the frameworks mentioned earlier, including principles derived from critical theories such as feminism and critical race theory. Recently, authors have introduced design frameworks for social justice from Design Justice (Ostrowski & Breazeal, 2022) or Humanitarian Engineering (Zhu et al., 2024) and principles from data feminism (Winkle et al., 2023), but clear guidance on how to do that in practice lacks (Lupetti et al., 2021), as most of the current frameworks take from existing principles and practices in other fields and need to be adapted in the ontological nature of HRI.

While current developments are crucial and invaluable, open questions remain. Are we perpetuating a system of oppression with our robot designs? Who are we othering in the HRI community and our design work? Who belongs, and who doesn't? (Kagedan, 2020).

Social justice projects and ethical approaches often stem from the privileged few (and offer little space for the ones oppressed by robots) and are tainted by ethics washing and shirking (Crawford et al., 2019; Floridi, 2019). The pressing question for those who want to engage in ethics and justice-oriented design is: who decides what justice is? Who benefits from the research and designs? Inclusive and just practices should prioritize understanding the historically complex and untangled relations between power and oppression and support the politics of belonging and care. We advocate focusing on ways of generating technology that favors radical participation (i.e., the design process conducted by the people most impacted by HRI design) and non-exploitive forms of existence between human and non-human agencies (Braidotti et al., 2016; de la Bellacasa, 2017). Ultimately, political reframing should focus on examining the exploitative and utilitarian roots of HRI and develop imaginaries and futures of justice, belonging, and care.

Epistemological and Methodological Reframe: Toward Transdisciplinarity and Methodological Openness

The premise of this book was the need for an epistemological broadening to extend how we make sense of and learn how to design robots and shape our relations with them (Chapter 1). We have discussed at length what this means and how to embrace intermediate-level knowledge. Considering the two reframes we discussed, an important facet going forward is to enable just social impact on top of disciplinary knowledge production.

While personal robots may not fully enter our everyday lives just yet, industry robots, robots to support logistics and delivery, and drones have been integrating into the fabric of society.

These developments call for rethinking who makes design decisions and integrating the reflection on societal implications in the research and design process. Mainly, HRI has been structurally defined by a minority of voices from the West-European and Japanese cultures (Seaborn et al., 2023), and societal participation in the development of robots is limited, raising the question of "epistemic control" and "epistemic justice" tied to colonization, capitalism, and slavery (Lewis et al., 2020). Robots' socio-technical complexity and implications are becoming apparent to researchers and practitioners alike (see Chapter 7), calling for value-driven and participatory modes of knowledge production.

To this end, many are considering transdisciplinary ways of knowing and practicing research when addressing challenges such as designing for the future of work (Zaga et al., 2024) or our cities.

Transdisciplinary research (Pohl & Hadorn, 2007) is a socially engaged way to practice knowledge production that is increasingly explored in design research from a variety of perspectives (Ozkaramanli et al., 2022; Thompson Klein, 2004; van der Bijl-Brouwer, 2022). The premise of engaging in transdisciplinarity is that complex societal challenges cannot be understood or tackled by a single academic discipline (Nicolescu & Ertas, 2008) or a particular group or community in isolation (Mobjörk, 2010). While many consider it a meaningless buzzword, it has been increasingly explored in practices to create value-driven collaborations between various actors in knowledge production. Transdisciplinarity requires challenging the status quo of scientific knowledge and practicing new ways of cooperation between scientists, designers, governments, companies, and citizens to integrate knowledge and bring about social transformations. It also requires methodological eclectism while engaging in participatory practices, which are typically facilitated by design methods and design futuring methods (Chapter 5) (Geenen et al., 2023; Matos-Castaño et al., 2020; Zaga & Lupetti, 2022). However, transdisciplinary research methods clash with typical HRI knowledge production modes. Transdisciplinary ways of working are opaque, situated in context, and emergent. It is an arduous process to detail, practice, and document. Plus, it does not conform with the typical way to assess rigor and validity as it is integrative. Therefore, understanding how to produce transdisciplinary knowledge and deal with values, pluralism, participation, and expertise integration beyond disciplines is yet to be defined in the HRI field. While transdisciplinarity remains a desirable epistemological reframe, much work lies ahead.

FINAL WORDS

Robots might be coming into our societies differently than we imagined. Still, the *résistance* is coming too. The scholars and practitioners featured in this book have crossed disciplinary lines while questioning HRI practices, producing invaluable methods, tools, and techniques. They offered reflections, methods, tools, and techniques that reject the strict disciplinary silos to explore how to design robots that work and matter for society (H. R. Lee et al., 2019; Lee et al., 2021; Vallor, 2016; Winkle et al., 2023; Zhu et al., 2024).

The ten chapters of this book are an inspiring testimony of how the field is growing, maturing, changing, and addressing its limitations. Still, questions remain even for those joining the *résistance*. Are the robots researched in HRI leaving the lab yet (Jung & Hinds, 2018)? Can we fully commit as a field to generating a positive societal and environmental impact (Crawford, 2021)? How can we do so, considering the politics of HRI (Zaga & Lupetti, 2022)? How is HRI confronting the fact that our work mainly comes from WEIRD (western, educated, industrialized, and democratic) countries (Seaborn et al., 2023)?

While we cannot answer these questions now, we see that HRI is engaging in a welcomed (re)evolutionary broadening. We hope the book can serve as a compass for current and future transformation. We hope it will impact our ways of working in HRI, contributing to widespread reflexivity, questioning, and curiosity. Finally, we hope to stimulate a transformative process beyond disciplinary divides by providing scholars, students, and practitioners with a window into possible epistemological, ontological, methodological, and political reframes.

In a world that is becoming increasingly complex, where technology is becoming increasingly a form of power and oppression (Crawford, 2021; Zaga & Lupetti, 2022) rather than a tool for human and societal flourishing, we believe that assessing, critiquing, and transforming our practice has become imperative to designing the futures we want to live in. This book, we hope, will inspire readers to develop robotic futures we would like to live in.

REFERENCES

Arista, N., Costanza-Chock, S., Ghazavi, V., & Kite, S. (2021). *Against reduction: Designing a human future with machines.* MIT Press.

Bartneck, C., Yogeeswaran, K., Ser, Q. M., Woodward, G., Sparrow, R., Wang, S., & Eyssel, F. (2018, February). Robots and racism. In *Proceedings of the 2018 ACM/IEEE international conference on human-robot interaction* (pp. 196–204).

Birhane, A., & van Dijk, J. (2020, February). Robot rights? Let's talk about human welfare instead. In *Proceedings of the AAAI/ACM conference on AI, ethics, and society* (pp. 207–213).

Braidotti, R., Disch, L., Hawkesworth, M., et al. (2016). *Posthuman feminist theory.* https://dspace.library.uu.nl/handle/1874/386619

Cila, N., Smit, I., Giaccardi, E., & Kröse, B. (2017, May). Products as agents: Metaphors for designing the products of the IoT age. In *Proceedings of the 2017 CHI conference on human factors in computing systems* (pp. 448–459).

Cila, N., Zaga, C., & Lupetti, M. L. (2021, June). Learning from robotic artefacts: A quest for strong concepts in human-robot interaction. In *Proceedings of the 2021 ACM designing interactive systems conference* (pp. 1356–1365).

Crawford, K. (2021). *The atlas of AI: Power, politics, and the planetary costs of artificial intelligence.* Yale University Press.

Crawford, K., Dobbe, R., Dryer, T., Fried, G., Green, B., Kaziunas, E., … & Whittaker, M. (2019). AI now 2019 Report. New York: AI Now Institute. *Search in.*

de La Bellacasa, M. P. (2017). *Matters of care: Speculative ethics in more than human worlds* (Vol. 41). U of Minnesota Press.

Dorst, K. (2011). The core of 'design thinking' and its application. *Design studies, 32*(6), 521–532.

Drenik, G. (2022). *Why robots are taking over the world – And that's a good thing.* Forbes. https://www.forbes.com/sites/garydrenik/2022/12/09/why-robots-are-taking-over-the-worldand-thats-a-good-thing/

Escobar, A. (2018). *Designs for the pluriverse: Radical interdependence, autonomy, and the making of worlds.* Duke University Press.

Eyssel, F., & Hegel, F. (2012). (s)he's got the look: Gender stereotyping of robots 1. *Journal of Applied Social Psychology, 42*(9), 2213–2230.

Floridi, L. (2019). Translating principles into practices of digital ethics: Five risks of being unethical. *Philosophy & Technology, 32*(2), 185–193.

Forlano, L. (2017). Posthumanism and design. *She Ji: The Journal of Design, Economics, and Innovation, 3*(1), 16–29.

Frauenberger, C. (2019). Entanglement HCI the next wave? *ACM Transactions on Computer-Human Interaction (TOCHI), 27*(1), 1–27.

Geenen, A., Matos Castaño, J., & van der Voort, M. (2023). The potential of smart city controversies to foster civic engagement, ethical reflection and alternative imaginaries. In *Rethinking technology and engineering: dialogues across disciplines and geographies* (pp. 143–155). Springer International Publishing.

Gero, J., & Milovanovic, J. (2023). The situatedness of design concepts: empirical evidence from design teams in engineering. *Proceedings of the Design Society, 3*, 3503–3512.

Giaccardi, E., & Redström, J. (2020). Technology and more-than-human design. *Design Issues, 36*(4), 33–44.

Hawthorne, J. (2007). Cartesian dualism. *Persons: Human and divine* (pp. 87–98).

Hoffman, G. (2019). Anki, Jibo, and Kuri: What we can learn from social robots that didn't make it. *IEEE Spectrum, 1*(05), 2019.

Hou, Y. T. Y., Cheon, E., & Jung, M. F. (2024, March). Power in human-robot interaction. In *Proceedings of the 2024 ACM/IEEE international conference on human-robot interaction* (pp. 269–282).

Jung, M., & Hinds, P. (2018). Robots in the wild: A time for more robust theories of human-robot interaction. *ACM Transactions on Human-Robot Interaction (THRI)*, *7*(1), 1–5.

Kagedan, A. L. (2020). *The politics of othering in the United States and Canada*. Springer International Publishing.

Klein, J. T. (2004). Prospects for transdisciplinarity. *Futures*, *36*(4), 515–526.

Kolko, J. (2010). Sensemaking and framing: A theoretical reflection on perspective in design synthesis.

Latour, B., & Weibel, P. (2005). *Making things public*. https://philpapers.org/rec/LATMTP

Lee, H. R., Cheon, E., De Graaf, M., Alves-Oliveira, P., Zaga, C., & Young, J. (2019, March). Robots for social good: exploring critical design for HRI. In *2019 14th ACM/IEEE international conference on human-robot interaction (HRI)* (pp. 681–682). IEEE.

Lee, H. R., Cheon, E., Lim, C., & Fischer, K. (2022, March). Configuring humans: What roles humans play in HRI research. In *2022 17th ACM/IEEE international conference on human-robot interaction (HRI)* (pp. 478–492). IEEE.

Lee, H. R., Sun, F., Iqbal, T., & Roberts, B. (2023, March). Reimagining robots for dementia: From robots for care-receivers/giver to robots for carepartners. In *Proceedings of the 2023 ACM/IEEE international conference on human-robot interaction* (pp. 475–484).

Lee, M., Noortman, R., Zaga, C., Starke, A., Huisman, G., & Andersen, K. (2021, May). Conversational futures: Emancipating conversational interactions for futures worth wanting. In *Proceedings of the 2021 CHI conference on human factors in computing systems* (pp. 1–13).

Lewis, J. E., Abdilla, A., Arista, N., Baker, K., Benesiinaabandan, S., Brown, M., … & Whaanga, H. (2020). Indigenous protocol and artificial intelligence position paper.

Lupetti, M. L., & Van Mechelen, M. (2022, March). Promoting children's critical thinking towards robotics through robot deception. In *2022 17th ACM/IEEE international conference on human-robot interaction (HRI)* (pp. 588–597). IEEE.

Lupetti, M. L., Zaga, C., & Cila, N. (2021, March). Designerly ways of knowing in HRI: Broadening the scope of design-oriented HRI through the concept of intermediate-level knowledge. In *Proceedings of the 2021 ACM/IEEE international conference on human-robot interaction* (pp. 389–398).

Luria, M., Oden Choi, J., Karp, R. G., Zimmerman, J., & Forlizzi, J. (2020, July). Robotic futures: Learning about personally-owned agents through performance. In *Proceedings of the 2020 ACM designing interactive systems conference* (pp. 165–177).

Matos-Castaño, J., Geenen, A., & van der Voort, M. (2020). The role of participatory design activities in supporting sense-making in the smart city.

Mobjörk, M. (2010). Consulting versus participatory transdisciplinarity: A refined classification of transdisciplinary research. *Futures*, *42*(8), 866–873.

Nicolescu, B., & Ertas, A. (2008). Transdisciplinary theory and practice. *USA, TheATLAS*.

Noel, L. A., Ruiz, A., van Amstel, F. M., Udoewa, V., Verma, N., Botchway, N. K., … & Agrawal, S. (2023). Pluriversal futures for design education. *She Ji: The Journal of Design, Economics, and Innovation*, *9*(2), 179–196.

Orduña, N. (2021, March 19). *Why robots won't steal your job. Harvard Business Review*. https://hbr.org/2021/03/why-robots-wont-steal-your-job

Ostrowski, A. K., & Breazeal, C. (2022, March). Design justice for robot design and policy making. In *2022 17th ACM/IEEE international conference on human-robot interaction (HRI)* (pp. 1170–1172). IEEE.

Ozkaramanli, D., Zaga, C., Cila, N., Visscher, K., & van der Voort, M. (2022). Design methods and transdisciplinary practices.

Perugia, G., Guidi, S., Bicchi, M., & Parlangeli, O. (2022, March). The shape of our bias: Perceived age and gender in the humanoid robots of the abot database. In *2022 17th ACM/IEEE international conference on human-robot interaction (HRI)* (pp. 110–119). IEEE.

Perugia, G., & Lisy, D. (2023). Robot's gendering trouble: A scoping review of gendering humanoid robots and its effects on HRI. *International Journal of Social Robotics*, *15*(11), 1725–1753.

Pohl, C., & Hadorn, G. H. (2007). *Principles for designing transdisciplinary research* (pp. 36–40). oekom.

Richardson, A. (2024, February 16). *Elon Musk walks his "Spooky" Tesla humanoid robot around that "Will Be Able To Do Everything Better Than Us" — As Microsoft and BMW partner on a rival robot*. Yahoo Finance. https://shorturl.at/mNZ56

Savage, N. (2019). Perfectly practical in every way. *Nature*, *567*(7749), S38–S40.

Schön, D. A. (2017). *The reflective practitioner: How professionals think in action*. Routledge.

Schön, D. A., & DeSanctis, V. (1986). "The reflective practitioner: How professionals think in action". *Journal of Computing in Higher Education*, *34*(3), 29–30.

Seaborn, K., Barbareschi, G., & Chandra, S. (2023). Not only WEIRD but "uncanny"? A systematic review of diversity in human–robot interaction research. *International Journal of Social Robotics*, *15*(11), 1841–1870.

Spiel, K., Frauenberger, C., Keyes, O., & Fitzpatrick, G. (2019). Agency of autistic children in technology research—A critical literature review. *ACM Transactions on Computer-Human Interaction (TOCHI)*, *26*(6), 1–40.

Strait, M., Ramos, A. S., Contreras, V., & Garcia, N. (2018, August). Robots racialized in the likeness of marginalized social identities are subject to greater dehumanization than those racialized as white. In *2018 27th IEEE international symposium on robot and human interactive communication (RO-MAN)* (pp. 452–457). IEEE.

Tanqueray, L., Paulsson, T., Zhong, M., Larsson, S., & Castellano, G. (2022, March). Gender fairness in social robotics: Exploring a future care of peripartum depression. In *2022 17th ACM/IEEE international conference on human-robot interaction (HRI)* (pp. 598–607). IEEE.

Toso, F., Zaga, C., den Haan, R. J., & Aizenberg, E. (2023). Do not believe the hype: Critically discussing the role and pedagogical implication of generative AI in human-centred and transdisciplinary design education. In *7th international conference for design education researchers 2023*.

Vallor, S. (2016). *Technology and the virtues: A philosophical guide to a future worth wanting*. Oxford University Press.

Van Amstel, F. M. (2023). Decolonizing design research. In *The Routledge companion to design research* (pp. 64–74). Routledge.

van der Bijl-Brouwer, M. (2022). Design, one piece of the puzzle: A conceptual and practical perspective on transdisciplinary design.

Verbeek, P. P. (2005). *What things do: Philosophical reflections on technology, agency, and design*. Penn State Press.

Wakkary, R. (2021). *Things we could design: For more than human-centered worlds*. MIT Press.

Walsh, G., Foss, E., Yip, J., & Druin, A. (2013, April). FACIT PD: a framework for analysis and creation of intergenerational techniques for participatory design. In *Proceedings of the SIGCHI conference on human factors in computing systems* (pp. 2893–2902).

Weiss, A., & Spiel, K. (2022). Robots beyond science fiction: Mutual learning in human–robot interaction on the way to participatory approaches. *AI & Society*, *37*(2), 501–515.

Wiltse, H. (2020). Human-Technology-Human Relations, or Politics by Other Means. In *Philosophy of Human-Technology Relations Conference 2020, 4–7 November, 2020*.

Winkle, K., McMillan, D., Arnelid, M., Harrison, K., Balaam, M., Johnson, E., & Leite, I. (2023, March). Feminist human-robot interaction: Disentangling power, principles and practice for better, more ethical HRI. In *Proceedings of the 2023 ACM/IEEE international conference on human-robot interaction* (pp. 72–82).

Winkle, K., Melsión, G. I., McMillan, D., & Leite, I. (2021, March). Boosting robot credibility and challenging gender norms in responding to abusive behaviour: A case for feminist robots. In *Companion of the 2021 ACM/IEEE international conference on human-robot interaction* (pp. 29–37).

Zaga, C., Lupetti, M. L., Forster, D., Murray-Rust, D., Prendergast, M., & Abbink, D. (2024, March). First international workshop on worker-robot relationships: Exploring trans-disciplinarity for the future of work with robots. In *Companion of the 2024 ACM/IEEE international conference on human-robot interaction* (pp. 1367–1369).

Zaga, C., & Lupetti, M. L. (2022). Diversity equity and inclusion in embodied AI: Reflecting on and re-imagining our future with embodied AI. *research.utwente.nl.* https://research.utwente.nl/files/285680270/DEI4EAIBOOKLET_WEB_SINGLEPAGES.pdf

Zhu, Y., Wen, R., & Williams, T. (2024, March). Robots for Social Justice (R4SJ): Toward a more equitable practice of human-robot interaction. In *Proceedings of the ACM/IEEE international conference on human-robot interaction (HRI).*

Index

Pages in *italics* refer to figures.

A

ACM code of conduct, 148
ACM/IEEE International Conference on Human-Robot Interaction, 162
Adelino robot, 9, 23–25
adversarial design, 104, 120–122, 149
Al-Jazari, 2
Alves-Oliveira, P., 103, 111–113
Amazon Alexa, 2, 76
Amazon's Echo, 119
American Nurses Association Study, 143
annotated portfolios, 165
anthropomorphism, 3, 79, 175
artifacts, 114, 116, 133–134, 149–150
  contestational, 104, 120–122
  design, 33, 132, 163–165
  generating, 102–104, 171
  intelligent, 89
  multitude of, 10, 32
  robotics, 7–8, 113, 154, 164–166, 172
  semantics of, 6
  technological, 149, 172
artificial intelligence (AI), 2, 71, 117
Auger, J. H., 104, 117–119, 150
automata, 100–101
automation, 11, 14, 57, 61, 76, 89, 101, 121–122, 170
AvantSatie game, 23–25, *24*

B

Bak, Sure, 26–28
Ball, A., 66
Barad, K., 29
Barendregt, W., 164
Bartneck, C., 175
Benjamin, W., 117
Bicchi, M., 134
big data, 119
Black in Robotics (BiR), 150
Bleeker, M., 10, 29–31
Bødker, M., 40, 58–60
Boll, S., 71
Boon, B., 70
Borchers, J., 71
Börjesson, P., 164
Borning, A., 46

C

Cagiltay, B., 32–34
Carter, E. J., 43
Cartesiam dualism, 174
Castellano, G., 172
Chairbot, 15–16
Chang, W. L., 39
Cheon, E., 103, 105–107, 172
Child–Computer Interaction community, 164
Chinese traditional culture, 127–128
Chu, Vivian, 141
Cila, N., 70–97, 161–167, 172
Clarkson, P. J., 6
Coeckelbergh, M., 71
collective imaginary, 1
colors, 6–7, 22, 132–133, 140
concept-driven designs, 155
Contreras, V., 175
conversational user interfaces (CUI), 2
Copenhagen Business School (CBS), 59, *59*
Cossin, I., 41, 52–54
COVID-19, 141
Crilly, N., 6
critical perspectives in HRI, 148
  critical design, 149
  critical research, 150
  critical robotics as design research program, 150–152
    marginal and norm-creative perspectives, 153–154
    moving beyond user requirements, 152, 154–155
    problematization, 152–153
    role of designer/positionality, 155–156
  ethics and technology, 149
  invitation to contribute, 156–157
    future critical robotics research, 157
  questions about design and use of robots in society, 149
  speculative design and design fiction, 150
"Critical Robotics exploring a new paradigm" workshop, 148
Cross, N., 163
Cukier, K., 119
curiosity(ies), 58, 73, 80, 89, 91, 97, 101, 116, 142, 171, 177

D

decentering the human, 71
decision-making, 41, 47, *50*, 83
Degrees of Freedom (DoF), 20
Delft University of Technology, 75
Dennett's theory of intentionality, 80
design; *see also* designing robots
in HRI work, *see* designerly HRI work
knowledge, 152–153, 161, 166, 172
methods, 3, 42, 153–155
of personal robots, 132; *see also* personal robot design
intentional material choice, Blossom robot, 135, *136–138*
process, 19, 21–22, 39, 41–42, 47, 52, 74, 102, 123, 143–144, 152, 154, 162–164, 171, 173, 176
research
Research for Design, 162
Research into Design, 163
of robotic artifacts, 7–8
of robotic devices, 132
and robotic innovation, 4
robot unboxing experiences, 33
role in HRI research, 161
technology, 43, 100
theory, 163–165
designerly HRI work
epistemology of, 163
first dichotomy
research *vs*. practice, 162–164
second dichotomy
theory *vs*. instance, 164–165
third dichotomy
academy *vs*. industry, 166–167
designerly inquiry, 161
Design Fiction, 103–105, 114–116
designing artificial agents
animation
AvantSatie game, *24*
ERIK technique, 23–25
neutral, warm and cold expressive postures, 25
iterative usability tests, 23
soul to an embodiment, 23
bodystorming, 26–28
in Parcel Delivery Center, *27*
scenarios of indoor delivery robots, 26–28
ARC Brain, 28
enacted robots, 28
dramaturgy for devices
speculative encounters with supermarket robots, 29–31
development of PAL Tiago robotos, 29
mixed-reality setup, *30*, 30
puppeteer, 30–31
toward performativity, 29
minimalism
the greeting machine
basic geometric shapes, 22
minimal metaphor minimal movement, 21–22
prototypes and mechanisms, *21*
robot's appearance design, 20
robot's mechanism and morphology design, 20
robot's movement design, 20
multi-modal expressivity
adjustable in robot-assisted therapy, *18*, 18–19
human–robot collaboration, 17
robot's perceived lifelikeness, 17
ready made prototyping, 14
from hoverboards to mobile robots, 14–16
Trash barrel robot, *15*, 15–16
isolate the reaction, 14
readymades, 14
robotization design of everyday things, 8
design strategies, 11–12
PopupBot, *12*
collaboration, 13
modularization, 12–13
stereotypical ideas/imagination of, 8
worldbuilding
as design research method, 32
offer context for design, 32
social robot for children, *33*, 33–34
designing for social embeddedness
breaching experiments, 40
Incidentally Co-present Persons, 58
CBS, 59
ethnomethodology, 60
provocations, 59
light-weight method, 58
membership classification, 58–60
co-design
design process, 42
design project, 55
feature of, 55
a multi-colored hexagon, *56*
process of mutual learning, 55
robot programming, 57
robot training sessions, 57
summer school research study, 57
collaborative map making
practical aspect of mapping, 49
reframing assistive robots with older adults, 49–51, *50*
HRI research practices and frameworks, 51
precise methodology, 51

end-user programming
 in aviation manufacturing
  automation, 61
  Franka Emika robot, 63
  interface, 61, *62*
  NASA ULI project, 62
 as low-code/no-code programming, 61
 methods
  program representations, 61
 visual programming, 61
 Wizard menus, 61
installations and performances
 Diamandini, *65*
  interactive humanoid robot, 64
  movement, 65–66
  sonification of, 66
  Sound, 64, 66
  Touch, 64, 66
 implicit communication modes, 64
 physicality, 64
pretotyping, 41
 assistive robots, 52
 concept of, 52
 goal of, 52
 low-tech living lab, 54
 Pepper robots, 54
 robot prototypes
  Anubis–non-humanoid actuated skeleton, 54
  Nao–a small size anthropomorphic robot, 54
  Romeo-a cardboard representation of a humanoid robot, 54
 Romeo2
  formal prototype, 52–53, *53*
  Intention Scenarios, 54
  low-tech prototypes, 53–54
robot value mapping
 adoption of, 46
 robot code of conduct
  aesthetic design guidelines, 47
  robot behaviors, 48
  robotic product teams, 48
 VSD, 46
user enactments
 apology and option strategies, 45
 application of
  robot's behavior, 43
 recovery strategies, 43
 scenarios, 43–44
 service literature, 45
 service robot storyboard, *44*
 structured, 43
designing human-robot ecologies
 aesthetic of friction
  controlled level of friction, 83
  Keymoment, 84–85
   with a bike and car key, *84*
  Pleasurable Troublemakers, 83
 conversations with agents, 71
  as decentering technique, 74
  interactions of devices, 74
  more-than-human perspective, 74
  workshop "in conversation with robots", *75*, 75–76
 Objects with Intent framework
  embedding, 81
  framing, 72
  a robot ball for physical play, 81–82
   Children's interactions, 82
   impression of Fizzy rolling around, *81*
  transformation, 80
 para-functionality approach
  Roomba+Clips project, 92–94
   accessory kit, *93*
  use of "functional" objects, 92
 playfulness approach
  promote curiosity and exploration, 89
  self-directed and open-ended learning, 89
  Woodie, a playful robot, 89–91, *90*
 relationality
  Lichtsuchende, society of robots, 95–97, *96*
  relational frameworks, 95
 symbiosis, 72
  in field of autonomous robotics, 86
  human-dependent social robot/weak robot, 86
  Swiss Army Knife approach, 86
  weak robots for symbiotic relations with human
   iBones robot, *87*, 88
   Sociable Trash Box, 87–88
 Techno-Mimesis
  being human and becoming a cleaning robot, *78*
  experiences of, 79
  prostheses, 77
  robotic superpowers, 77, 79
  with a shopping robot, 78–79
  vs. Wizard-of-Oz technique, 77
Designing Interactive Systems (DIS) in 2020, 75
designing robots
 extreme interaction design, 147
 Moxi robot/project; *see also* Moxi robot
  emergence of robot–worker relationship, 141–143, *142*
  from the lab to the real world, 140–141
  principles for robot design
   continuously keep the team aligned, 144–146
   rely on in-depth design research, 143–144
   take advantage of real-world deployment, 146
 Diana, C., 140–147

Diligent Robotics, 141, 143
direct-drive Brushless DC (BLDC) motors, 16
DiSalvo, C., 43–45, 120
Dörrenbächer, J., 72, 77–79
Drone Chi, human–drone interaction experience, 108–110, *109*
Dunne, A., 92–93
Durrant-Whyte, H., 40

E

Effective Human–Robot Teaming To Advance Aviation Manufacturing, 62
electrical impedance tomography, 66
emotions, 6–7
end-user programming (EUP), 61–63
Erel, H., 20–22
Eriksson, E., 164
ethnographic experiential futures (EXF), 104, 123–125
Eyssel, F., 175

F

Fallman, D., 166–167
feedback, 14, 28, 39–40, 128, 141, 146
Fish-Bird exhibit in 2005, 40
Fitzpatrick, G., 175
Forlizzi, J., 38, 40, 43–45, 163, 173
Foundation of Responsible Robotics, 150
Franka Emika robot, 63
Frauenberger, C., 70, 175
Frayling, Christopher, 162–163
Friedman, B., 46
function-oriented modularization, 13
futuristic autobiographies (FABs), 103, 105–107

G

Gallo, D., 26–28
Gamboa, M., 103, 108–110, 148–157
Garcia, N., 175
Garfinkel, H., 58
Giaccardi, E., 70–71, 74–76, 163
Gonzalez, Irene, 41, 46–48
Grasso, M. A., 26–28
Gross, D., 126
Guidi, S., 175

H

Harding, S., 155–156
Harraway, D., 153
Hassenzahl, M., 71, 78, 83–85
Haunted Desk, 14
Hegel, F., 175
Heidegger, Martin, 70
Hello Robot, 116
Hoffman, Guy, 132–138
Hoggenmueller, M., 89–91
Hou, Y. T. Y., 172
HoverBot, robot platform, 8, 14–16
human-computer interaction (HCI), 26, 70, 149–150, 161, 164–166, 172, 174–175
humanities, 3, 70, 102, 161, 171
human-like/zoomorphic entities, 7
human-made objects, 132
Human-Robot and Human-Agent Interaction, 124
human–robot interaction (HRI), 2–4, 31, 38, 40, 71, 76, 102, 134, 141
  application in, 66
  community, 3, 73, 165, 167, 170
  Conference, 164
  critical robotics in, *see* critical perspectives in HRI
  design activities, 41, 155, 163
  designerly contributions to field of, *see* designerly HRI work
  nonverbal modality in, 134
  questions, 27, 171
  reframing
    epistemological and methodological
      transdisciplinarity and methodological openness, 176–177
    learning from and, 172–173
    ontological
      connecting relationality with pluriversality, 174–175
    political
      justice, belonging, and care, 175–176
  research, 166
    and practice, 60
    and robot design practices, 73
  and robotics
    in public sector and policy making, 3
  robot movement in, 65
  status quo of, 171
  theories, 165, 167
  understanding of, 40
human–robot relations, 71–72, 170–172, 174–175

I

iBones robot, *87*, 88
Ihde, D., 71
Incidentally Co-present Persons (InCoPs), 58
industrialization, 101
Industrial Revolution, 101
industrial robotic arms, 6–7
industrial robots, 6–7
Iqbal, T., 173
iRobot, 16, 92

J

Jacobs, Jan, 41, 46–48
Johnson, Khari, 141
Jung, M. F., 163, 172
Ju, W., 8, 14–16

K

Kahn, P., 46
Kang, D., 8, 11–13
Kaptelinin, V., 70
Karp, R. G., 173
Keyes, O., 175
Kiesler, S., 43
Kim, Kahyeon, 26–28
Kinect Azure, 55
kinetic embellishment of motion, 63
Kwak, S. S., 8, 11–13

L

La Delfa, J., 103, 108–110
Lakatos, Imre, 151
Lang, Fritz, 101
large language models (LLMs), 170, 172
Larsson, S., 172
Laschke, M., 72, 83–85
Lee, C. P., 32–34
Lee, H. R., 38, 49–51, 173
Lee, M. K., 43–45
Lee, W. Y., 163
Leite, I., 172
Lely, 166–167
Liezi, traditional Taoist Chinese text, 2
Lisy, D., 173
Ljungblad, Sara, 174
Lloyd, P., 162
Löffler, D., 78
Lupetti, M. L., 1–4, 6–34, 104, 120–122, 162–163, 173
Luria, M., 43–45, 104, 123–125, 163, 173

M

Mandel, I., 8, 14–16
mapping values, 41
marginal perspectives, 152–153, 156–157
Mayer-Schönberger, V., 119
McMillan, D., 172
mechanical age, 100–101
Mechanical Love, 39
Melsión, G. I., 172
metaphors, 20, 100, 103, 111–113
Mi, Haipeng, 126–128
mobile umbrellabot, 15
Moja robots, 127–128, *127*
Moultrie, J., 6
movement, 7–9, 20–22
Moxi robot
- 4-foot-tall mobile robot, 140
- Diligent's Moxi Healthcare Robot, *142*
- hospital robot
  - to help nurses, 141–143
- interactive product design principles, 143–146
- launch, 143
- minimal architecture, 141

multi-modal expressivity, 9, 17–19
Murray-Rust, D., 73, 95–97
Mutlu, B., 32–34, 38, 43–45
My Robot Gets Me: How Social Design Can Make New Products More Human, *142*, 144

N

Neato vacuum robot, 16
neutrality ideal, 155
New Metaphors Toolkit, 112–113
Nicenboim, I., 71, 74–76
non-verbal communication, 9, 17, 128
non-verbal modalities, 17
norm-creative perspectives, 153–157

O

Objects with Intent (OwI) framework, 80–82, *81*
Ocnarescu, I., 41, 52–54
Oden Choi, J., 173
Okada, M., 72, 86–88
Oltman, D., 43
orchestration, 13

P

PAL Tiago robots, 29
Parlangeli, O., 175
participatory design, 55, 108, 154, 173
Paulsson, T., 172
personal robot design
- auditory expression, 133–135
- material choice of, 132
  - Blossom robot, 135
    - hand-crocheted wool cover and handcrafted wooden ears, *136*
    - interior, *137*
    - wooden mechanical parts, *138*
  - tactile interaction, 133
  - time and place, 134–135
  - visual aesthetics, 132–133

Perugia, G., 173, 175
phone's joystick, 63
Pierce, J., 92–94
PopupBot, robotic pop-up, 8
positionality/role of designer, 155–156

problematization, 152–153
product semantics, 7–8
Pütten, A. R. V. D., 71

Q

questioning, 119, 122, 149, 156, 171, 177

R

Raby, F., 92–93
Ramos, A. S., 175
real prediction machines (RPMs), 117–119, *118*
Rebaudengo, Simone, 99–101
Redström, J., 103, 114–116
reflexivity, 171, 177
Reig, S., 43–45
relational databases, 95
relational movements, 40
research-design convergence, 163
Research through Design (RtD), 163–164, 172
Research Through Design Conference (RTD2019), 75
RGB-D image, 63
Ribeiro, Tiago, 23–25
Roberts, B., 173
Robinson, F. A., 66
robot behavior, 9, 17, 27, 29, 31, 46, 48, 57–58, 60–62
Robot Film Festival, 116
robotic imaginaries
  adversarial design
    contestational artifacts, 120
    practices, 120
    Steering Stories project, *121*, 121–122
  blending traditions
    an ongoing cultural creation, 126
    cultural coordinate, 126
    Moja robotic Chinese orchestra, *127*, 127–128
    in traditional culture, 127
  Design Fiction
    creative and speculative approach, 114
    prototypes, 114–116
    teacher of algorithms, *115*, 115–116
    "what if" scenarios, 114
  ethnographic experiential futures
    design research method, 123
    letters from the future, 124–125, *124*
    stages of design processes, 123
  futuristic autobiographies, 103, *106*
    configuring the user: "robots have needs too", 105–107
    diegetic user, 105
    grounded theory, 107
    value elicitation tool, 105
  metaphors
    create shared meanings, 111
    for human–robot interaction, 111–113
      inspired by tumor as relational metaphor, *112*
      three-stage design exploration, 113
  Soma Design, 103
    felt experiences, 108
    first-person perspective, 108
    Somatic drones, 108–110
      Drone Chi, human–drone interaction experience, 108–110, *109*
  speculative design, 117
    real prediction machine, 117–119, *118*
    strategies, 117
robotics, 1–2
  artifacts, 7–8, 113, 164–166, 172
  history of, 7
  and society
    mutually shaping, 39–40; *see also* designing for social embeddedness
Robot Institute of America, 1
robotization, 7–8
robots'
  20th century concept, 100
  Adelino, 23–25
  agency and intentions of, 72
  alterity of, 71
  appearance design, 20
  as artificial agents, 1–2
  cleaning, *78*, 78–79
  configuration of, 6
  Cosmo (a toy robot), 76
  CUIs, 2
  Degrees of Freedom, 20
  design, 8–10, 38–42, 47, 73, 126, 143, 147, 162, 166
    personal, *see* personal robot design
  emotional responses, 6–7
  Franka Emika, 63
  group working sessions, 57
  and HRI; *see also* human–robot interaction
    vision for design, *4*
  human-like, 11
  hype and frenzy, 170
  images of, 101
  interactions, 32–34
    between humans, 70
  kinetic language for, 65–66
  mechanism and morphology design, 20
  as mediating agents, 175
  reachability of, 63
  Real Prediction Machines, 104, 117–119, *118*
  seal-like Paro, 39
  Simon, 140
  social context of use, 39

socially embedded, 39, 41; *see also* designing for social embeddedness
socio-technical complexity, 176
stereotypes and norms, 8
term, 1
therapeutic robot *Paro*, 6
user evaluations of, 38
vacuum cleaning, 16
Woodie, a playful robot, 89–91, *90*
robot-to-human communication, 9, 17
Rosenthal-von Der Pütten, A., 58–60
Rozendaal, M. C., 29–31, 70, 72, 80–82
Rybski, P., 43
Rye, D., 40

S

Šabanovicì, S., 38–66
Savoia, A., 52
scenario prototyping, 54
Scheding, S., 40
science fiction, 148–149
Senft, E., 42, 61–63
Ser, Q. M., 175
Silvera-Tawil, D., 64
situated action, 40
Smart Design, 140, 143
Sociable Trash Box, 86
social justice, 175–176
Socially Guided Machine Learning (SG-ML), 140
The Socially Intelligent Machines (SIM) Lab, 140–141
social robot, 148, 151
Somatic drones, 108–110, *109*
Sparrow, R., 175
speed dating study, 44
Spiel, K., 175
Srinivasa, S., 43
Stappers, P. J., 163
Stegner, L., 43
Steinfeld, A., 43
Stolterman, E., 164, 166–167
Strait, M., 175
Suchman, L. A., 40
Sun, F., 173
Su, N. M., 103, 105–107

T

Talos, 2
Tan, H., 38
Tanqueray, L., 172
therapeutic robot *Paro*, 6
Thomaz, Andrea, 140–141
Thrun, S., 1
Toasterbot, 14
Torgersson, O., 164
transdisciplinary research, 176–177
Trash barrel robot, *15*, 15–16
Turtlebot, 16

U

Umbrello, S., 46
user enactments, 43–45
UX designers, 166

V

value-grounded robot behaviors, 46
Value-Sensitive Design (VSD), 46
Van de Poel, I., 46
Van Mechelen, M., 153
Velonaki, M., 40, 64–66
Verbeek, P. P., 71, 73, 149
Video-based HRI (VHRI), 54
visual aesthetics, 132–133
Vitra Design Museum, 116

W

Wakkary, R., 70
Wang, J. Z., 43
Wang, S., 175
weak objectivism, 155
WEIRD (western, educated, industrialized, and democratic), 41, 101, 177
Wiberg, M., 164
Willerslev, R., 77
Winkle, K., 55–57, 172
Wizard of Oz
capabilities, 43
simulations, 59
technique, 54, 77
Woodie, a playful robot, 89–91, *90*
Woodward, G., 175
Wulf, V., 71

Y

Yogeeswaran, K., 175

Z

Zaga, C., 162, 170–177
Zhong, M., 172
Zimmerman, J., 43–45, 163, 173
Zuckerman, O., 9, 20–22

Printed in the United States
by Baker & Taylor Publisher Services

Printed in the United States
by Baker & Taylor Publisher Services